Springer
Tokyo
Berlin
Heidelberg
New York
Hong Kong
London
Milan
Paris

K. Takeuchi, R.D. Brown, I. Washitani,
A. Tsunekawa, M. Yokohari (Eds.)

Satoyama

The Traditional Rural Landscape of Japan

With 85 Figures

Springer

Kazuhiko Takeuchi, Dr. Agr.
Professor
Laboratory of Landscape Ecology and Planning
Graduate School of Agricultural and Life Sciences, The University of Tokyo
1-1-1 Yayoi, Bunkyo-ku, Tokyo 113-8657, Japan

Robert D. Brown, Ph.D.
Professor
School of Environmental Design and Rural Development, University of Guelph
Guelph, Ontario N1G2W1, Canada

Izumi Washitani, Dr. Sci.
Professor
Laboratory of Conservation Ecology
Graduate School of Agricultural and Life Sciences, The University of Tokyo
1-1-1 Yayoi, Bunkyo-ku, Tokyo 113-8657, Japan

Atsushi Tsunekawa, Dr. Agr.
Associate Professor
Laboratory of Landscape Ecology and Planning
Graduate School of Agricultural and Life Sciences, The University of Tokyo
1-1-1 Yayoi, Bunkyo-ku, Tokyo 113-8657, Japan

Makoto Yokohari, Dr. Agr.
Associate Professor
Institute of Policy and Planning Sciences, University of Tsukuba
1-1-1 Tennodai, Tsukuba, Ibaraki 305-8573, Japan

Cover: The watercolor reproduced on the front cover is by Chizuru Kamon, depicting a traditional *satoyama* landscape created by the coexistence between nature and humans.

ISBN 4-431-00007-0 Springer-Verlag Tokyo Berlin Heidelberg New York

Library of Congress Cataloging-in-Publication Data applied for.

This book is based on the Japanese original, K. Takeuchi, I. Washitani, A. Tsunekawa (Eds.), *Satoyama no Kankyogaku*, University of Tokyo Press, 2001.

Printed on acid-free paper

Typesetting: Camera-ready by the editors and authors

SPIN: 10887006

Preface

The traditional term *satoyama* has recently come back into common usage in Japan. Environmentally conscious individuals use satoyamas to symbolize nature that requires management. The term *satoyama*, however, does not have a specific definition and this has sometimes caused confusion when scientific discussions are attempted. In this book, satoyamas are defined as secondary woodlands and grasslands adjacent to human settlements. The more general term *satoyama landscape* describes the broader area of secondary nature, including satoyamas, as well as cultivated lands, human settlements, and wetlands.

This book aims to describe satoyama landscapes from the scientific perspective, as well as discuss directions for conservation from three different points of view: scientific, citizen participation, and political.

The main focus of this book is on the nature of satoyama landscapes. The assertion of this book is that the secondary nature in satoyama landscapes is a consequence of a coexistence between nature and humans. In order to conserve satoyama environments, adequate land management is as essential as physical land protection.

Satoyama landscapes are suffering from pressures associated with development and, to make matters worse, the few remaining satoyamas are being neglected. It is essential in satoyama conservation to determine how to continue with human intervention and, at the same time, to act so as to protect the remaining physical land area. Attempts are made in this book to reveal biodiversity transformation through human intervention in order to propose a direction for satoyama management.

However, sustainability of satoyamas cannot be discussed without consideration of the integrity of the broader satoyama landscape. The fuel revolution has deprived satoyama landscapes of one of their most important functions. Cultivated lands, human settlements, and wetlands are also suffering from land abandonment and the destruction of secondary nature due to urban development. In this book, the condition of satoyama landscapes is reviewed through a chronology of land use. The importance of nature conservation is evaluated and a systems approach to conserving nature is discussed. One of the conclusions resulting from this review is that there is a need to redevelop sustainable societies based on bioresources.

Concerned citizens play a major role in satoyama landscape conservation. Satoyama landscapes are considered not only as a surviving natural environment, but also as *the nature of the good old days*, the Japanese idyllic landscape. The recognition of the need for adequate management can motivate citizen participation. When one considers that, in the past, only farmers conducted satoyama landscape management, the situation has changed quite dramatically. Both the participants and the techniques have changed. This book argues for the need to support citizen movements, separate from bureaucracy, but with appropriate cooperation and recognition of their roles and abilities.

The 21st century is said to be the century of the environment. We need to change our relationship with nature, which has seen a sad segregation develop through modernization, and seek coexistence. People need to coordinate with global perspectives and work within their own region to create living environments that are in harmony with nature. This way of thinking could be a touchstone to fuse global and local cultures together. I expect that this book will be a step toward those attempts for the future.

This book was edited based on the results of a research group on the establishment of a strategy for the conservation of satoyama landscapes, aiming to achieve integrated and planning-oriented conservation of satoyama landscapes, which was supported by the Nippon Life Insurance Foundation for 2 years in 2000 and 2001. The Japanese edition of this book was published by the University of Tokyo Press in November 2001.

The international edition of the book was edited based on the Japanese edition and necessary modifications were made. Particularly, for international readers, a new chapter describing the global view of the satoyama landscape was added in order to clarify the uniqueness and similarity of satoyama landscapes to coppice woodland utilization throughout the world. I believe that this book will contribute to the promotion of worldwide comparative studies on this topic by the expected readers, such as researchers, citizens, and administrative officers, as well as non-governmental organizations (NGOs) that are actively involved in conservation activities.

I am deeply grateful to Dr. Amal Kar, Dr. Satoru Okubo, Ms. Rui Seguchi, Mr. Marco Amati, Ms. Yuki Sampei, and Ms. Kaoru Ichikawa who helped the editing works. I would like to express sincere thanks to the Nippon Life Insurance Foundation, The University of Tokyo Press, and Springer-Verlag Tokyo, who helped us publish this book. I also would like to express our gratitude to the Japan Society for the Promotion of Science for providing us with a Grant-in-Aid for Publication of Scientific Results (No. 145337) for the fiscal year 2002.

Kazuhiko Takeuchi
Editor-in-Chief

Contents

Contributors

Name of Author (Chapter and Section)

Yoshimi Asou (5.2)
Sakuragaoka Park Coppice Volunteers
3-6-9-402 Takaban, Meguro-ku, Tokyo 152-0004, Japan
E-mail: ranocchi@h3.dion.ne.jp

Atsuki Azuma (4.4)
Faculty of Agriculture
Iwate University
3-18-8 Ueda, Morioka, Iwate 020-8550, Japan
E-mail: azuma@iwate-u.ac.jp

Tsutomu Bessho (3.2)
Ohtori Consultants (Environmental Design Institute)
3-10-10 Minamisenba, Chuo-ku, Osaka 542-0081, Japan
E-mail: b_tsutomu@hotmail.com

Robert D. Brown, Professor, Ph.D. (1)
School of Environmental Design and Rural Development
University of Guelph
Guelph, Ontario N1G2W1, Canada
E-mail: rbrown@la.uoguelph.ca

Kenji Iiyama, Professor, Dr. Agr. (6.1)
Director, Asian Natural Environmental Science Center
The University of Tokyo
1-1-1 Yayoi, Bunkyo-ku, Tokyo 113-8657, Japan
E-mail: uakenji@mail.ecc.u-tokyo.ac.jp

Yoshiko Kitagawa (5.3)
Laboratory of Landscape Ecology and Planning
Graduate School of Agricultural and Life Sciences
The University of Tokyo
1-1-1 Yayoi, Bunkyo-ku, Tokyo 113-8657, Japan
E-mail: ayoshiko@mail.ecc.u-tokyo.ac.jp

Yoshinobu Kitamura, Professor (2.4, 7.2)
Faculty of Law
Sophia University
7-1 Kioicho, Chiyoda-ku, Tokyo 102-8554, Japan
E-mail: kitamu-y@sophia.ac.jp

Noboru Kuramoto, Associate Professor, Dr. Agr. (2.3, 4.1, 5.2)
Faculty of Agriculture, Meiji University
1-1-1 Higashimita, Tama-ku, Kawasaki, Kanagawa 214-8571, Japan
E-mail: kura@isc.meiji.ac.jp

Hideharu Kurita (3.4)
National Institute for Rural Engineering
2-1-6 Kan-non-dai, Tsukuba, Ibaraki 305-8609, Japan
E-mail: hkurita@nkk.affrc.go.jp

Junko Morimoto, Dr. Agr. (3.3)
Graduate School of Media and Governance
Keio University
5322 Endo, Fujisawa, Kanagawa 252-8520, Japan
E-mail: junkomo@sfc.keio.ac.jp

Yukihiro Morimoto, Professor, Dr. Agr. (3.3, 4.3)
Laboratory of Landscape Ecology and Planning
Graduate School of Global Environmental Studies
Kyoto University
Kitasirakawa-Oiwakecho, Sakyo-ku, Kyoto 606-8502, Japan
E-mail: ymo@kais.kyoto-u.ac.jp

Shigetoshi Nakagawa (5.1, 6.2, 6.3)
Natural Environmental Conservation Center, Kanagawa Prefecture
657 Nanasawa, Atsugi, Kanagawa 243-0121, Japan
E-mail: junen-ab-bp@nifty.com

Youichi Sonoda (4.1)
Doctoral Course
Graduate School of Agriculture
Meiji University
1-1-1 Higashimita, Tama-ku, Kawasaki, Kanagawa 214-8571, Japan
E-mail: sonogonia@yahoo.co.jp

Kazuhiko Takeuchi, Professor, Dr. Agr. (2.1, 7.3)
Laboratory of Landscape Ecology and Planning
Graduate School of Agricultural and Life Sciences
The University of Tokyo
1-1-1 Yayoi, Bunkyo-ku, Tokyo 113-8657, Japan
E-mail: atake@mail.ecc.u-tokyo.ac.jp

Atsushi Tsunekawa, Associate Professor, Dr. Agr. (3.1, 3.2, 7.1)
Laboratory of Landscape Ecology and Planning
Graduate School of Agricultural and Life Sciences
The University of Tokyo
1-1-1 Yayoi, Bunkyo-ku, Tokyo 113-8657, Japan
E-mail: atsushi@mail.ecc.u-tokyo.ac.jp

Izumi Washitani, Professor, Dr. Sci. (2.2, 4.2, 5.4)
Laboratory of Conservation Ecology
Graduate School of Agricultural and Life Sciences
The University of Tokyo
1-1-1 Yayoi, Bunkyo-ku, Tokyo 113-8657, Japan
E-mail: awashi@mail.ecc.u-tokyo.ac.jp

Makoto Yokohari, Associate Professor, Dr. Agr. (1, 3.4)
Institute of Policy and Planning Sciences
University of Tsukuba
1-1-1 Tennodai, Tsukuba, Ibaraki 305-8573, Japan
E-mail: myoko@sk.tsukuba.ac.jp

1 Ideological Contribution of Satoyamas

R. D. Brown and M. Yokohari

1.1 Introduction

The 20th century has seen dramatic changes in the way that humans use the landscapes of the world. People are now living in and affecting virtually every environment on earth. There will soon be few truly natural areas left. While it is crucial that we preserve the natural areas, an even more important issue for the 21st century will be to determine how people can live in harmony with the natural environment.

This book considers solutions for the future by looking back. Over the centuries, a pattern of land use emerged in rural Japan that was sustainable over a long period of time. The satoyama landscape system was a harmonious relationship between humans and nature. The nature of the satoyama landscape was managed, and both humans and nature benefited from this management. As human populations continue to grow, and people move into previously unsettled areas, we need successful prototypes like satoyamas as patterns for the development of settlements that will allow both people and nature to survive.

Habitat modification is not a uniquely human trait; in fact, it is quite common in nature. Virtually every organism modifies its environment to its advantage. Examples of natural landscape modifications include everything from African termites building impenetrable mounds, to black walnut trees (*Juglans nigra*) warding off competition by exuding *juglone*, which is poisonous to other plants, to the beaver (*Castor canadensis*) of North America, which cuts down forests of trees and builds dams that flood large tracts of land. Managed landscapes can have a great deal of biodiversity and add an important dimension to the landscape mosaic. In contrast, many human-dominated environments are almost devoid of nature. This is not a prototype for sustainability. Rather, we should be seeking to integrate humans and nature in a manner that is beneficial to both.

The traditional satoyama landscape of Japan did exactly that. The system was devised, through trial and error, as a means of surviving on small mountainous islands in a monsoon climate. Japan is a land of natural disasters. Earthquakes, volcanic eruptions, torrential rains, and typhoons are all normal occurrences. The natural environment of Japan developed under these conditions. As a consequence, many of the native species are naturally tolerant of disturbance, human or otherwise.

The heavy rains of Japan collect quickly in steep mountain streams and make a short, violent dash to the sea. Flooding and property damage are inevitable without careful planning and management. The water management system that was devised and used successfully throughout Japan to mitigate the effects of storms was the basis for a prototypical settlement pattern. The system was simple but very effective (see Figure 2.1). The hillsides were covered with managed woodlands and carefully terraced paddy fields. Human settlements were arranged linearly along the base of the hills. Immediately in front of the settlements were the vegetable gardens, and beyond that were the lowland paddy fields. Rainfall was buffered by the trees, so caused little erosion. The water was slowly but methodically passed from the woodlands and terraced paddies, through the villages to the vegetable gardens, then through paddy after paddy until it gently entered the local stream or river. All along the way, water provided an abundant habitat for wildlife and slowly percolated into the groundwater.

Materials were also cycled within the system. Leaves from the trees were composted and added to vegetable plots and paddy fields. Branches were used for fuel either directly or as charcoal. After trees reached a desired size they were harvested and the stumps were left to sprout. This coppice woodland was managed carefully so there was a constant supply of wood and the woodland always had an open canopy, allowing sunlight to penetrate and maximize the biomass production. Inputs to the system were limited to those provided by nature: precipitation and solar radiation. Outputs included sales of agricultural products and fuel to downstream urban settlements. It was a simple but very effective system that persisted for many generations.

The urban residents of Japan were also deriving a tremendous benefit from the satoyama landscape, perhaps without even realizing it. The most important of these benefits was flood protection. Massive flooding was the norm when heavy monsoon rains would fall on the steep mountains of Japan. Enormous volumes of water would have to traverse short, steep rivers that inevitably ran through urban areas located on the plains below. The water management system of the satoyama landscape mitigated this effect. Another major benefit was slope stabilization. Landslides and soil erosion could make travel and trade treacherous, but the managed satoyama woodlands stabilized the rural environment.

In the middle of the 20th century, however, several events served to disrupt the system. Fossil fuels become more readily available, easier to use, and less expensive than charcoal or wood. Chemical fertilizers became easier to acquire and more effective than composted leaves. Rural residents joined in the worldwide trend of moving into urban areas. Suddenly, the satoyama landscapes that had been central to Japanese life for so long were abandoned and nearly forgotten. Coppice woodlands became dense and overgrown. Understory plants disappeared in the heavy shade and the woodland floor started to erode. Carefully tended paddy field dikes crumbled, allowing water to move more quickly through the landscape, creating more soil erosion. Satoyamas were the forgotten landscape of Japan.

After several decades of decline, the satoyama landscape has only recently begun to be rediscovered. Scientists have begun to measure land use change over time and to determine the effect on everything from biodiversity to landscape stability. Local residents are beginning to realize the natural and cultural value of the satoyama landscapes. Grass-roots community groups are forming to manage the landscapes in traditional ways and give them new life. This resurgence in interest has led to the production of this book. It was originally produced only in Japanese, but we soon realized that there are lessons here that can benefit others outside Japan. Concepts from satoyama landscapes are applicable more broadly. Perhaps we can learn from systems of the past and apply them to the future. This book is intended to document and share information about coppice woodlands and their associated landscape systems in Japan so that others might learn from it.

1.2 International Perspective on Coppice Woodlands

1.2.1 Experiences in Other Countries

The satoyama landscape system revolves around the coppice woodland. There is evidence that woodlands have been managed this way since ancient times, not only in Japan, but also in many other parts of the world. Coppices continue to be areas of considerable interest internationally, with recent studies investigating aspects of coppices in the United Kingdom (e.g. Buckley 1992), Sweden (e.g. Rydberg 2000), France (e.g. Logli and Joffe 2001), the Czech Republic (Konvicka and Kuras 1999), Italy (e.g. Amorini et al. 2001; Cutini 2001), USA (Anderson 1999), Malawi (Abbot and Lowore 1999), Kenya (Kennedy 1998), Korea (Hong et al. 1993), and New Zealand (Sims et al. 2001), to name but a few. The concept has currency in both developed and developing countries, as well as in both traditional and contemporary applications.

The UK has a long history of coppice woodlands that has been going on for hundreds, and in some cases thousands, of years (Buckley 1992). The practice of coppicing trees has created unique ecosystems that, where there is less dead wood, results in a more open canopy, allowing in more solar radiation and, in many cases, a higher diversity of plants and animal species than unmanaged woods. As in Japan, the coppices of the UK are no longer commercially viable and are currently of interest primarily for aesthetic and scientific reasons. Many are being destroyed either through neglect, clearing, or replanting to other species (Buckley 1992). The problems facing UK coppices are similar to the Japanese situation: little or no cutting has taken place in half a century; there is a lack of skilled labor; and management by volunteers does not provide a long-term solution.

Since 1945, the removal of a management regimen in coppices in the UK has altered their ecology (Peterken 1992). Many coppices have been cleared, mostly for agricultural activities, so that distances between remaining coppices have increased. Many of the ancient hedges that connected so many coppice woodlots have been removed to enlarge agricultural fields, so connectivity between woodlots has generally decreased. Streams have been altered to improve drainage, meadows and pastures have been reduced or removed, large amounts of agricultural chemicals have been introduced into the environment, and coppices that have survived destruction have been abandoned without management (Peterken 1992).

As in Japan, people in other countries are seeking answers to the coppice question: what should be done with woodlands that were managed for a long time and have now been abandoned? Some efforts have focused on reviving traditional methods of management, whereas others have sought new uses for coppices that lead to economic reasons for their survival.

1.2.2 Use of Coppice Wood for Fuel

In many parts of the developing world, wood is still the main source of domestic energy. One of the key problems in sustainable land use in Africa, for example, is the on-going need for fuel wood and the consequent destruction of native forests. The use of coppices for producing a continuous and renewable source of energy shows considerable promise in countries such as Malawi (Abbot and Lowore 1999) and Kenya (Kennedy 1998).

Other recent investigations have focused on the selection of species of trees for coppicing that will provide the best source of fuel (Senelwa and Sims 1999; Sims et al. 2001), and on energy and carbon budgets from coppices (e.g. Matthews 2001).

1.2.3 Biodiversity

There is currently interest in the UK in the effects of coppicing on the biodiversity in the coppice woodland. There seems to be a growing consensus on how coppices should be managed to provide maximum biodiversity with the increased understanding of how coppices affect plants, birds, mammals, and insects (Buckley 1992). Studies have been done comparing coppice management with natural disturbances (Evans and Barkham 1992), the effects of coppice activities on perennial ground flora (Barkham 1992), woodland breeding birds (Fuller 1992), small mammals (Gurnell at al. 1992), and butterflies (Warren and Thomas 1992). A recent study in the Czech Republic suggested that an endangered butterfly is threatened with extinction due to loss of coppice habitat. These butterflies require the sunny, open environment of the coppice and are disappearing as abandoned coppices become closed-canopy, even-aged woodlands (Konvicka and Kuras 1999). Coppices have been found to provide functions such as pest control and wildlife habitat (Sage 1998). Production is not compromised and few pesticides were required, and, if appropriately located, coppices have the potential to increase biodiversity in many farmland situations.

1.2.4 Contemporary Applications

Coppices are being actively investigated as possible tools for the ecological restoration and recovery of contaminated lands. Recent studies have investigated the use of coppice woodlots for extracting contaminants such as landfill leachate in the USA (Stephens et al. 2000) and radiocesium in Belarus and Western European sites (Vandenhove et al. 2001). Studies in Sweden have investigated coppice treatment of wastewater (Rosenqvist et al. 1997) and investigations have considered the use of coppices as carbon dioxide mitigation measures in the UK (Scholes 1998) and Belgium (Dubuission and Sintzoff 1998).

1.2.5 Developing Versus Developed Countries

There is limited scholarly literature currently available in either Japanese or English that describes the historical and contemporary status of traditional coppice landscapes in other Asian countries. Hong et al. (1993) suggest that the situation in Korea is very similar to that of Japan. Rural landscape patterns in China, as described by Forman (1995), are also similar to the satoyama landscapes of Japan. They are described as a relationship between humans and their natural environment, where woodlands are maintained on potentially erodible slopes and water is carefully managed as it flows slowly through the settlement.

However, the situation in other Asian countries seems to be very different. In developed countries, such as Japan, Korea, and, indeed, almost all the countries described so far in this chapter, the traditional managed landscapes are faced with the problem of abandonment and overgrowth, while in developing countries the situation appears to be the exact opposite, one of overutilization (e.g. Brookfield and Byron 1993). Both situations have occurred because of changes in economic and political contexts, and both lead to degradation of the environment. However, the overutilization of wooded landscapes is outside the scope of this book.

Studies of the satoyama landscapes of Japan must be considered in this broader context. The sections of this book cover many of the same topics that are of interest in other regions of the world, and hopefully the Japanese experience will provide an interesting and valuable dimension.

1.3 Contributions to the Book

The chapters and sections of this book have been written by many different individuals. The authors are a diverse group, with a wide range of expertise, but they have one thing in common: they study the landscape. Satoyama landscapes have been considered from the biological, physical, social, cultural, political, and economic perspectives.

Chapter 2 introduces the nature of satoyama landscapes and defines the concepts of *satoyama* and *satoyama landscape*. Chapter 2 also provides a prehistoric and historic context by discussing the factors that affected satoyama development over time. The discussion of the nature of satoyama landscapes considers not only renewable resource extraction and species diversity, but also grass-roots conservation efforts and government policies. This section sets the context for the discussions in the rest of the book.

Satoyama landscapes are then discussed at various temporal and spatial scales in Chapter 3. The changes in the landscape over time are illustrated, starting at the national scale, moving to a regional level, and, finally, discussing specific areas. Two important case studies are considered: the Kanto Plain and the Kyoto Basin in Kansai. The mechanisms behind the change in landscape over time are discussed in detail.

The details of the biodiversity of satoyama landscapes are considered in Chapter 4. Plants, birds, and insects, as well as ecological mechanisms that have affected ecosystem relationships over time, are discussed within the context of a coppice landscape. The effects of vegetation management on the diversity of plant and bird species are discussed, with a special emphasis on rare and endangered species.

Chapter 5 outlines in detail current approaches to the conservation of satoyama landscapes. Citizen movements are analyzed to explain how and why they emerged, and how they can be supported in the future. Approaches to the regeneration of satoyama landscapes are introduced.

The vast biological resources of satoyama landscapes are investigated in Chapter 6 for their possible use in contemporary Japanese society. Wood harvested from satoyamas is investigated as a potential building material, both as lumber and as processed wood products. The history of the use of satoyamas as an energy resource is discussed and future scenarios are considered. Finally, the potential of satoyamas for use in nature studies is investigated and the value of participatory involvement in active management is discussed.

The final chapter (Chapter 7) discusses the history of conservation strategies in Japan and reviews current approaches to management. Future scenarios are considered, including the use of satoyama landscapes for energy, raw materials, mitigation against global warming, and legal regulations. A proposal is put forward for a national policy for landscape planning and a long-term national policy for xxx.

1.4 Contributions of the Book

The Japanese have long been known as a culture willing to import ideas and incorporate them into their lifestyle and culture. However, Japan is also a land with much to share with the rest of the world. Unfortunately, the Japanese language provides a challenging barrier. Few non-natives can read Japanese easily and share Japan's valuable contributions with the scientific community. In addition, translation of text from Japanese to English is not easy. Not only are the written characters different, but the manner in which the concepts are presented is sometimes almost incomprehensible when direct translations are made.

However, in this case the translation was worth the effort. Information about a traditional system for a sustainable coexistence with the natural environment must be shared. Through countless generations of trial and error, the Japanese have learned how to live in harmony with their environment, and not just in the short term. The satoyama landscapes of Japan were a wonderful, sustainable land-use system that thrived for centuries, probably millennia.

The attraction of city life has drawn most Japanese people from the countryside, leaving a dwindling number of mostly elderly people to live on the land. These elderly people are increasingly unable to manage the rural landscape as it falls further and further into disrepair. The management of the satoyama landscapes is a critical link in a complex system of sustainable land use and without it the whole system will collapse. The challenge is to determine what components of this system need to survive and how to make their management relevant and applicable in today's world.

This book introduces many concepts, some very old, some very new. The authors and editors hope that the information will contribute to the establishment of sustainable landscapes, where human and non-human organisms can live together for the benefit of all.

2 The Nature of Satoyama Landscapes

2.1 Satoyama Landscapes as Managed Nature

K. Takeuchi

2.1.1 What is Satoyama?

To Japanese people, the term *satoyama* conjures up images of the idyllic rural landscape of fields and woodlands. However, it is surprising that this term only became popular within the past 40 years. The term *satoyama* was used as long ago as 1759 by a Kiso area assistant wood manager by the name of Hyouemon Teramachi, who described satoyamas in a book entitled *Miscellaneous Stories of Kiso Mountain*. He described satoyama as mountainous landscapes close to rural villages (Tokoro 1980). The person who revived the term in modern times was Tsunahide Shidei, a forest ecologist, who proposed the idea of satoyamas in the early 1960s. He later explained that this term is just a modification of *yamasato* (village in the mountains) to satoyama (mountains near the village) so that everybody can understand the meaning. Based on this idea came the concept of the satoyama as an agricultural woodland (Shidei 2000).

The term *coppice woodland* expresses the role of woodlands in traditional rural life. A coppice woodland is used to produce wood or charcoal through a management system that involves cutting every 10 years or so. Although the trees are cut, the roots and stumps are left intact. Sprouts emerge from the stump and create a new woodland. This woodland reproduction system is called coppice regeneration. These woodlands are also used every year to provide dead branches, leaves, and herbaceous plants to create compost for traditional agriculture. Most of these woodlands are composed of pines or mixed deciduous species, and are called secondary woodlands because they were formed artificially. However, after World War II, when chemical fertilizers and fossil fuels came into common use, the economic importance of coppice woodlands diminished.

Today, nature-loving citizens often use satoyama to symbolize remnant natural environments in Japan. Citizens have become more aware of the conservation of the natural environment because of the loss of nature due to rapid urbanization and environmental destruction in suburban areas. In the late 1960s, there was a surge of suburban development along with rapid economic growth in Japan. In particular, coppice woodlands on hillslopes and dissected uplands were cleared for large-scale residential development sites. The sites for such large-scale housing development in the suburbs, such as Senri New Town and Senboku New Town in Osaka, Kouzouji New Town in Nagoya, and Tama New Town in Tokyo, were all once satoyama landscapes (Tamura et al. 1983). To develop extensive housing sites in satoyama landscape areas, the hill ridges were cut and used to fill the valley bottom to create a flatland. As a consequence, the natural environment of the satoyama landscape was completely destroyed.

Subsequently, people developed an increasing awareness of nature conservation in their neighborhood, and housing construction methods have been changed to partly maintain slopes covered with woodlands and to make the best use of natural landscapes. Conservation of satoyama landscapes came to be undertaken at development sites. Moreover, the citizens who moved to these new towns started voluntarily managing the surrounding woodlands and rice paddies. These citizens were getting more and more interested in satoyama landscapes and the concept of conservation rapidly started to influence all of Japan. In the process, *satoyama* has become a popular word for describing the traditional Japanese rural landscape.

Today, the word *satoyama* is used in various contexts, but it generally indicates a natural environment that is being managed and, therefore, its basic element can be represented as secondary nature. Secondary nature is easily lost in the process of large-scale urban development but, on the other hand, if it remains untouched it will be thoroughly transformed by natural vegetation succession. In order to conserve satoyama landscapes, adequate management is essential, as has been observed in traditional agricultural activities. In other words, we can see the richness of the natural environment of satoyama landscapes not only as an original natural diversity, but also as a natural diversity enriched by human intervention.

There are, however, different interpretations as to what elements can be included in the category of secondary nature. There is no doubt about including coppice and pine woodlands in this category because they are seen as typical plant communities. Also, in the traditional Japanese agricultural system, areas for thatch grass collection were regarded as very important and, therefore, these areas, too, can be included in satoyama landscapes. The problem is whether we should also include farmland and settlements in the satoyama landscape category. Considering that the traditional rural landscape is formed with coppice, grasslands, farmlands, and settlements, these elements must all be included as a set of landscape elements.

2.1.2 Satoyama Versus Satoyama Landscape

Moriyama (1997) mentioned that the typical land use arrangement of woodlands (*yama*), farmlands (*nora*), and settlements (*mura*) could be observed in upland rural communities in the Kanto Plains. Here, the word *yama* is used to indicate woodlands and, therefore, this yama can be regarded as similar to satoyama. It might be inappropriate, however, to use the word *satoyama* for farmlands and settlements. In this book, we propose to follow the lead of Fukamachi et al. (2001) and Yokohari et al. (2001) and call the entire area *satoyama landscape*.

Figure 2.1 shows the concepts of satoyama and satoyama landscape used in this book and classifies these concepts by relating them to the landform and land use

(Yamamoto 2001). As shown in the figure, *satoyama landscape* indicates the rural landscape that is comprised of satoyama, farmlands, settlements, and reservoirs, because these elements were once strongly connected to each other through the agricultural land use system.

In December, 1994, the Japanese Basic Environment Plan was implemented and it fully recognized the importance of satoyama landscapes. In this plan, the coexistence between nature and humans was regarded to be one of the most important long-term goals. Also, in order to achieve the goal, the coexistence of nature and humans with high natural and social quality in national lands was proposed, and policy implementation corresponding to each natural area with different natural environments was required. This *natural area* includes mountainous areas, satoyama landscapes, flatlands, and seashore areas. *Satoyama landscape* was defined as the middle part between mountainous areas and flatlands.

The Basic Environment Plan (Environment Agency 1994) describes the satoyama landscape as an area that has considerable secondary nature and an area that allows wild animals and humans to live together. The natural environment of this area is created through human interventions such as farming and forestry, and it is what Japanese people have long imagined as their idyllic landscape. In order to conserve the coppices, rice paddies, and irrigation ponds of satoyama landscapes, some action by citizens and public support are necessary.

Here, *satoyama landscapes* include not only the mixed woodlands, but also the *yatsuda* (special type of paddies, see below) and the small rivers and artificial ponds used for irrigation. Mixed woodlands of satoyama that are very near rural settlements and distributed on mountains, hills, and uplands as major geomorphic sites are very important in the rural villages to be used for coppice woodlands or agri-

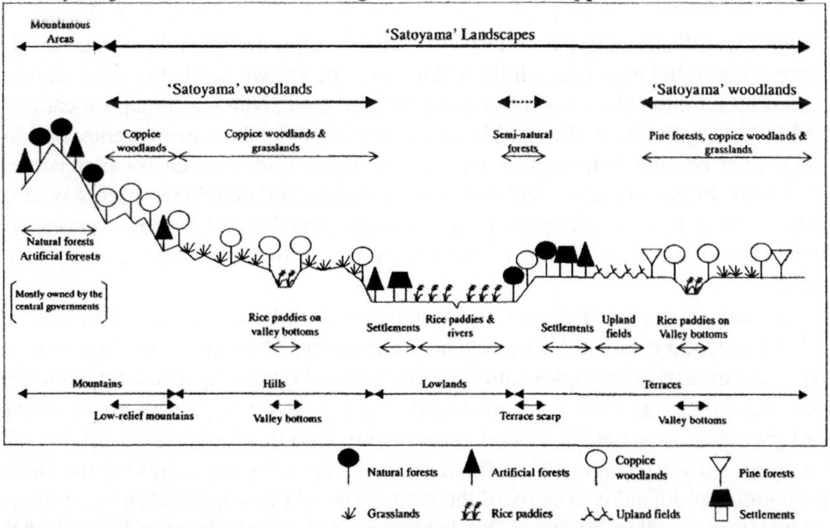

Fig. 2.1. Schematic representation of satoyama and satoyama landscapes (After Yamamoto 2001)

Fig. 2.2. Satoyama and yatsu distributed on the Kashima Uplands, Ibaraki Prefecture (left), and satoyama on the Hiki Hills, Saitama Prefecture (right). (Photographs by T. Okunuki)

cultural woodlands. On the other hand, narrow valley bottoms developed by dissecting low-relief mountains, hills, and uplands are known locally as either *yatsu* or *yato*. Further downstream is the alluvial lowland zone, which is a major location of paddy field agriculture. The paddy fields developed on the narrow valley bottoms use springs coming from the valley head. This type of rice paddies is called *yatsuda* or *yatoda*. In this book, we will define *yatsu* as lowland developed narrowly at the valley bottom and *yatsuda* as the paddy field developed on yatsu. However, in some regions, the local term of *yato* is popular, so only in this area will we use the term *yatoda*.

Yatsu developed in the low-relief mountains and hills are shaped like fingers on a hand extending into the slopes. The combination of reaches and lowlands are representative of the complex satoyama landscape (Figure 2.2). The terrace surfaces developed on uplands form a simple landform, but the terrace scarps are developed and presented as a site for natural or secondary woodland, which are followed by the dissected yatsu (Figure 2.2). The reason why these terrace scarps are the habitat for natural woodland is because of the narrowness of the woodland zone, so human maintenance is often difficult. This landscape is also considered to be part of the satoyama landscapes, which are composed of both the satoyama and the yatsu. Such satoyama landscapes are considered by people standing on the lowlands to be

Fig. 2.3. A yatsu landscape comfortably surrounded by satoyama woodlands

comfortably surrounded by nature, a nature that is very attractive in and of itself (see Figure 2.3).

Management of both satoyama and yatsu is also necessary to conserve the complementary wildlife habitats of plants and animals that live in the hills or in the uplands, and those that live in the wetlands. This contributes to maintaining the high biodiversity in the secondary nature. It must be noted that secondary nature has been maintained by the traditional farming activities both managing the satoyama and cultivating the yatsuda. Nowadays, as the role of the satoyama as coppice or agricultural woodland has decreased, it has also broken the tight connection with the yatsuda. Further, because of the decrease in the number of farmers and the low price of rice, many farmers have abandoned the both the yatsuda and the coppice woodlands.

The combination of satoyama and yatsuda is not only providing attractive landscapes for people, it is also a valuable habitat for wildlife. In Chapter 4, we describe the relationship between satoyama landscapes and yatsudas by using the example of the gray-faced buzzard (*Butastur indicus*). This bird stands at the peak of the ecosystem and its breeding grounds are in the satoyama landscape. During the breeding season, the gray-faced buzzards must provide plenty of food for their young. This buzzard, at the time of rice planting, when the paddies are flooded, catches frogs and snakes. Later in the season, when the paddies are dry, it becomes difficult to catch frogs and, instead, the buzzards hunt insects in the surrounding woodlands. When the buzzards are searching for food, they perch at the top of a

Fig. 2.4. Typical land-use pattern of Santome Sinden in Saitama Prefecture. There is a linear arrangement of settlements along roads, with upland fields and woodland behind

high tree, where they have a good view of the surrounding landscape and can view their prey. This bird has adapted to the spatial structure of the satoyama landscape and its seasonal changes.

There is another type of satoyama landscape that exists. This type is secondary woodland and grassland on flat uplands called *heichilin* or *flatland woodlands* and has been used to maintain agriculture on nutrient-poor uplands. On the Kanto Plain, where these uplands are widely distributed, these satoyama landscapes are common. *Santome Shinden* (newly developed agricultural fields) in Saitama Prefecture illustrate a typical example of this type of satoyama landscape. The land use pattern consists of settlements along the roads, upland fields extending back from individual homes, with flatland woodland beyond that (Figure 2.4). On the other hand, in the Shimofusa Uplands, with its rather wide flatlands, flatland and yatsu are alternately distributed. On this flatland, there are also flatland woodlands. The 1881 survey map, at a scale of 1 : 20 000, indicates that, in the Ushiku and Okami districts in Ibaraki Prefecture, the ratio of areas between woodlands and agricultural lands was approximately 2 : 1 (Sprague et al. 2000). This suggests that, for the agricultural activities at that time, twice as much woodland as agricultural land was needed.

However, such satoyama landscapes lost their relationship with society after the 1960s, when the satoyama landscape lost its function in agricultural practices. Therefore, such places became a target for development. Many of the flatland woodlands became sites of housing developments or industrial compounds that had come about as a result of factories moving out of city centers. The flatland woodlands also became dumping sites for industrial waste. There was a big issue of dioxin related to waste treatment in Tokoruzawa City in Saitama Prefecture. Compared with housing development with large-scale earthworks on the low-relief mountains or on the hills, landform transformation on the uplands is not so serious. However, in flatland woodlands on uplands it is also remarkable to see the decrease in secondary nature.

2.1.3 Understanding Nature from Two Perspectives

In this book we are distinguishing *satoyama* from *satoyama landscape* because we wanted to clarify the understanding of nature. The scientific field of landscape ecology describes the region as a series of ecosystems and provides principles for environmental conservation of such ecosystems (Troll 1971; Forman and Gordon 1986; Takeuchi 1991). In landscape ecology, there are two ways to view nature. One is to see the interrelation among natural elements, such as geology, geomorphology, soils, water balance, and vegetation; that is, to see it vertically. Satoyama is to be understood as a spatial unit, as a vertically homogeneous combination of such natural elements.

Let's take a typical example of a satoyama unit on uplands. The materials comprising the upland terraces are accumulated layers of sand, gravel, and clay. In the Kanto Plain area, these layers are often covered with volcanic ash. Andosols form on this type of landform. Such uplands, in general, are very dry and poor in nutrients. It is necessary to irrigate and to add nutrients to the soil if you want to use it for agriculture. Satoyama units on the uplands are also important as a coppice woodland for providing fuels and as an agricultural woodland for providing compost for agricultural lands. Even though the area was originally a broad-leaved evergreen forest, secondary pine or mixed deciduous woodlands had been maintained. Therefore, the satoyama as a unit is a combination of vertical natural elements modified by human interaction.

The other view of landscape ecology is the horizontal view, related to the heterogeneity of nature. This can be described by *catena*, or an understanding of nature as a sequence. The main element affecting this catena is landform. The plains surrounded by mountains and composed of hills, uplands, and lowlands are regularly distributed due to the geologic structure. In each geomorphic region, there are specified vertical relationships of natural elements, but such a natural sequence is contributing to the increase in regional landscape diversity. There is no doubt that, in these sequences, there are natural patterns or processes reflecting the interaction of geology, landform, soils, and water balance.

The relationship between satoyama and yatsuda shows the typical characteristics of a sequence(Figure 2.5). Two entirely different landscape units exist next to each

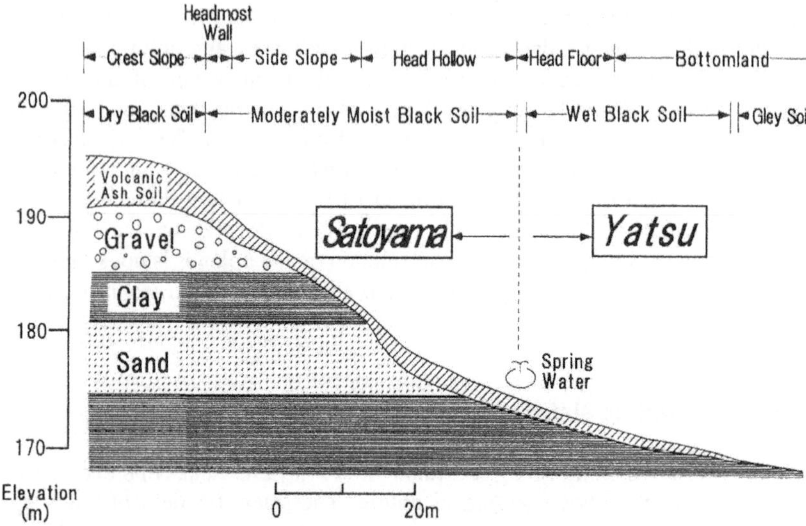

Fig. 2.5. Horizontal and vertical combination of natural elements in satoyama and yatsu in the Tama Hills (Matsui et al. 1990)

other and form a set of sequences of nature. More important is that mixed woodlands and yatsuda paddies that have existed in these two landscape units have been maintained by agricultural activities. Once we have considered the biodiversity created by the combination of these two units, it is necessary to find ways of maintaining both units in discussing the future direction of satoyama landscapes.

2.2 Satoyama Landscapes and Conservation Ecology

I. Washitani

2.2.1 History of Satoyama Landscapes

Nature that is apparent today is a consequence of human and natural processes that have taken place over thousands of years. Any natural area on the planet has been affected in some way, with very few exceptions, by human processes. Even in so-called virgin nature, some evidence of human interference can often be found. Of course, in some places human processes exceed that of natural processes and vice versa in others.

How have humans altered nature in the Japanese archipelago and when did this alteration begin? There is debate among archaeologists over when the first human settlement occurred in Japan. Despite scanty excavations of human bones (a result of the high acidity of the soil, which is unsuitable for bone preservation over long periods of time) archaeologists have discovered and analyzed ruins and human bones excavated from alkaline limestone soil. These samples suggest that humans first settled in Japan at least one million years ago.

Analysis of ruins and relics of the Paleolithic age reveal that ancient human lifestyles relied upon abundant natural resources from the surrounding mountains and oceans through fishing, hunting, and gathering. Evidence suggests that agriculture was first practiced in Japan in the Jomon period during the New Stone Age, a time when agriculture began around the world. Agricultural production apparently did not make a great contribution to food supplies during this time. It is not hard to imagine that agricultural practices were unnecessary when there was an abundant natural supply of food.

In addition to human remains, many animal bones, such as deer and wild boar, have been discovered among the ruins. Analysis of fat that remained on human bones excavated in Hokkaido indicated that, during the Jomon period, people ate more animal than fish protein. It can be assumed from this finding that hunting was an important factor in food management. Besides this discovery, ruins have revealed that grapes, raspberries, and other fruit were also important food resources. Wild grapes and raspberries grow in open areas or gaps, implying that woodland disturbance had occurred and gaps of larger areas had developed.

Such vegetation gaps were possibly a consequence of natural phenomena, such as wild fire caused by volcanic activity, earthquakes, and landslides due to typhoon activities (Washitani 2001a). The vegetation gaps may also be evidence that humans actively burned woodland to create areas that were desirable for hunting and gathering berries. It is believed that settlers of the Old Stone Age in Europe and North America burned woodland areas to deliberately create vegetation gaps. The act of burning woodlands in order to create vegetation gaps for hunting or gathering fields was possibly one of the oldest methods of vegetation management that humans applied to the land.

Because most of the paddy fields that have been discovered in Japan so far are from the Yayoi period, it is considered that this was the time when rice paddy field cultivation began. It is assumed that, prior to this period, rice was cultivated as a field crop rather than a paddy field crop. Paddy fields from 6000 years ago have been discovered in the Mimi Kaizuka Kita Ruin in Chitose, Hokkaido. Analysis of phytoliths (plant cell fossils) indicates early cultivation of plants of the millet family. A mass of deer bone fragments was also found. Leek phytoliths from 4000 years ago were found in the Gomyouda Ruin in Tonbara, Shimane, and phytoliths of Japanese millet from 3000 years ago were found in the Itaya Ruin of Tonbara, Shimane. In addition, phytoliths from the rice plant family were found in a 6000-year-old stratum in the Asanehana Ruin, Okayama (although this finding lacks reliability because samples have been too old for scientific proof). The oldest reliable relic in connection with rice in Japan was discovered in the Himezasahara Ruin in Bikan, Okayama. An excavated earthen vessel contained not only phytoliths, but also some fiber similar to straw and husks.

There is a possibility that crop fields in the Old Stone Age were managed by fire. Gathering, hunting, and fishing also seem to have played an important role in food supply, even after genuine rice cultivation started during the Yayoi period.

2.2.2 Genealogy of Living Through the Use of Plant Resources

Paddy fields from 2300 years ago were discovered in the Ikeshima/Hukumanji Ruin in Osaka. The remaining structure of the paddy fields was found in layers, which indicates that there were more than 20 previous paddy fields beneath the current one. This is attributed to the continuous creation of paddy fields after repeated destruction of the fields by floods. Including this case, more than 1000 remaining structures of paddy fields have been discovered across the country. Floods carry organic matter from mountain woodlands to alluvial plains and fertilize the soil. Volcanic ash soil, which holds nutrients well, can be made even more fertile through the mixing of organic matter. This is a desirable condition for permanent rice cultivation. Volcanic activity, and erosion and sedimentation by rivers are key factors in creating fertile soils that are ideal for crop cultivation. There is no conclusive evidence as yet to indicate whether settlers intentionally mixed organic matter with the soil for cultivation purposes.

How did the settlers in the Jomon period utilize plant resources other than for food supply? Some ruins provide clues. For example, in the Ofuna C Ruin, in Minami Kayabe, Hokkaido, residential remains were discovered dating back some 4000–5000 years. There were 1000 remaining structures, including oval buildings of 2.4 m in depth and 11 m in length. Large amounts of ash, consisting of *Miscunthus sinensis* grass and bamboo grass, were found in fire pits in the remains of pit dwellings, which implies the use of *M. sinensis* grass for fuel. *M. sinensis* grass was also used as thatching material for building structures. *M. sinensis* grass, which has been benefiting people in many ways as fodder or a building material until very recently, must have been a very important plant resource for centuries.

Processed chestnut wood has been discovered in the Sakuramachi Ruin, Koyabe, Toyama. The wood, thought to be 4000 years old, was used as structural material for soaking nuts in water to remove bitterness. It is believed to have been buried through flooding and has been preserved underground. Signs of processing were found on 50 of 200 excavated wood posts. In addition, bunches of bark and reed were found to have been used for constructing roofs and walls. Fragments of clothing from 6000 years ago, weaved with plants of the nettle family, were discovered in the Torihama Kaizuka Ruin, Sanpou, Fukui. This clothing is similar to clothing called *Angin* made from *Boehmaria nivea* in the Tohoku area and worn until recently.

The lifestyles of people in the Jomon and Yayoi periods revealed by archeological discoveries suggest a prototype of the satoyama landscape life system that is based on utilizing plant resources from the surroundings, as well as producing agricultural products.

2.2.3 The Abundant Nature of Volcanic Islands in a Monsoon Climate

The ecosystems in the Japanese islands are a consequence of a combination of three processes:

1. The natural processes of climate, soil, and topography of volcanic bands that arise where continental plates and oceanic plates collide;
2. Biological processes that have adapted to the surrounding environment; and
3. Human processes since settlement in the Neolithic Age.

In short, the ecosystem in the Japanese islands is a consequence of a complex combination of monsoon climate, mountainous landscapes, human life, and culture.

As mentioned earlier, fertile soil on volcanic ash is one of the conditions that guarantees agricultural prosperity. This type of soil is desirable for both paddy fields and dry fields. In addition to these soil conditions, adequate amounts of precipitation from monsoon climates enable the vigorous growth of crops. Flooding caused by concentrated precipitation during the rainy season or typhoons supplies organic matter, which adds further fertilizers to the soil.

At least for the past few thousand years, settlers on the islands have been living by creating crop fields by burning woodlands, expanding paddy fields by managing water, gathering wood and grass for fertilizer, fuel, or construction materials, and fishing in rivers and lakes. As a result, some species have died out, while others were forced to reduce their habitat. On the other hand, other species have taken advantage of the new conditions and have expanded their habitat. Many of the successful species were those that had adapted to natural disturbances, such as wild fire caused by volcanic activity, landslides caused by earthquakes, and wind disaster and flooding caused by typhoons. The ability of these species to deal with natural disturbances meant they were better able to deal with later human disturbances.

Wetland wildlife, such as species of frogs and dragonflies, has greatly extended its habitat into paddy fields and detention ponds. The combination of such artificial wetlands and adjacent woodlands (for purposes of plant resource supply) has greatly benefited amphibious animals and insects, such as the dragonfly.

Since the New Stone Age, human processes on the islands have influenced the ecosystems in many ways. However, these human activities were not intense enough to do irreparable harm to the biodiversity or ecological integrity essential to human existence. The abundant Japanese idyllic landscape that existed until some decades ago suggests that the processes were in balance. The idyllic landscape provided the Japanese people with visible and invisible blessings through the four seasons, including fruits, mushrooms, herbs, and cogon grass, all of which have been utilized by people since the Old Stone Age.

Thus, although the landscape of Japan is occasionally subject to disasters, it has been supporting the prosperity of people since the Paleolithic Age.

2.2.4 Sustainable Systems for Renewable Resource Extraction

Relevant human intervention enables sustainable production and a lifestyle based on natural resources. In this regard, satoyama landscapes might provide a prototypical sustainable system.

Satoyama landscapes contain nature that is maintained through its use and management by the local human population. Grasslands and diverse woodlands were created around residential areas and water bodies, thus allowing easy access to fertilizer and water for cultivation, fodder for cattle, wood, cogon grass, and bamboo for construction material, firewood and pine leaves for fuel, vine and bamboo for daily necessities, and so forth.

Such a system was based on traditional lifestyles of cultivating rice crops, using organic compost, living under thatched structures, and creating warmth from firewood and pine needles. As mentioned previously, a prototype can be dated back to the Jomon period, when people started to make the transition from gathering, hunting, and fishing to a more settled life.

Relevant disturbances and stresses are indispensable conditions in maintaining biodiversity in satoyama environments. Disturbance, by definition, is destruction of plant material, whereas stress is an environmental effect that constrains the photosynthesis process. If these constraints did not exist, competitive plants would monopolize available resources and plant communities with a limited number of species would be established. Human demand for resources rooted in traditional life have been modifying, through disturbance, plant communities of a limited number of species into coexisting systems of diverse species.

Of course, in satoyama landscapes the principle of environmental carrying capacity applies without exception. Overuse of resources and spatial limitations result in less biodiversity, as well as resource exhaustion, such as the mountain landscapes in the Kansai region, which have been denuded as a result of the overuse of woodland resources over time. Resource extraction should not exceed carrying capacity when managing natural resources in a sustainable manner.

In many of the areas that have successfully managed the sustainable use of plant resources, the use of satoyama landscapes as common land has been established. The Edo period has recently attracted attention as an era of agriculture. This was a time when local autonomy allowed the use of land as commons. Several considerations of sustainability can be identified within the practice of satoyama landscape management, which improved over time as a result of trial and error. For example, the harvesting time of grass and wood that was in high demand as fertilizer was strictly controlled and they were harvested so as to not prevent regeneration the following year. This intensive type of management ensured sustainable resource extraction.

Plants accumulate organic matter in their roots and stems while enabling photosynthetic production through the development of leaves and shoots. The rate of production of new plant material depends not only on the plant species, but also on seasonal conditions. Perennial plants, such as *Miscunthus sinensis* grass, recover above-ground substance rapidly in spring using organic products accumulated underground. When seasons change and the organic production in the leaves increases, the product is passed underground for accumulation. At that time, photosynthetic activation is restricted by activation of accumulation. This is known as the *restriction of the sink.* If the above-ground portion of a plant is removed at the wrong time, organic production for the year will decrease markedly. However, if this material is removed during the restriction of the sink, productivity will not be affected and growth in the following year can be guaranteed with enough underground accumulation of organic substances.

Harvesting oak after a certain amount of organic accumulation has been achieved, but during the time when it is still young enough for coppicing, enables the greatest productivity in the long term. People learned through experience the best time to harvest organic materials. Traditional relationships between people and satoyama landscapes were based on techniques and knowledge of sustainable natural management.

2.2.5 Species Diversity in Satoyama Landscapes

According to Whittaker (1960), there are three factors affecting species diversity, namely a diversity, ß diversity, and γ diversity. a diversity represents diversity within a habitat, ß diversity represents diversity of the habitat itself, and γ diversity represents regional diversity. In other words, a diversity is species diversity found within homogeneous habitat patches, ß diversity is diversity of patches that results from different settlers in different patches (Figure 2.6), whereas γ diversity is regional diversity (e.g. woodland and grassland species that occur in different regions and contain different fauna due to historical and biogeographical differences).

The most important factors affecting species diversity in satoyama landscapes are disturbance (destruction of part of the plant substance itself) and stress (prevention of photosynthesis bioproduction processes) by human extraction of plant resources. Disturbance and stress promote a diversity. In general, competition results in low species diversity areas because the occupation of competitive species eliminates uncompetitive species. In a satoyama landscape, where adequate disturbance and stress are present, competitive species are forced to yield to uncompetitive species to some extent, which enables diverse species to survive. This also provides food and habitat, both of which also positively affect the diversity of fauna.

As mentioned previously, nature in Japan accommodates human intervention relatively easily. The wide range of natural disturbances in Japan resulted in the evolution of plants that were readily able to adapt to environmental changes, such as cutting, mowing, and burning due to human intervention. Oaks that sprout from the base after being cut down and *Miscanthus sinensis* and *Miscanthus sacchariflorus* grass or common reeds that grow significantly despite mowing are good examples of this.

Miscanthus sinensis is common in Japanese grasslands. It has been one of the most important plant resources for people since the Jomon period. Cattle were fed *M. sinensis* and the resulting manure was used to fertilize crop fields. *Miscanthus*

Fig. 2.6. Satoyama landscapes have a mosaic of habitats

sinensis and *M. sacchariflorus* were also used to thatch roofs. There are indications that these grass may have also been used as fuel during the Jomon period.

The satoyama landscape system is a mosaic of sustainably maintained environments. It is a result of previous generations of people managing nature to support their livelihood. Such environmental mosaics are very conducive for biodiversity. Higher ß diversity is achieved by different types of species residing in different habitats. Environmental mosaics can support species that require more than two different habitats for survival. Satoyama landscapes that contain water bodies and adjacent woodlands are ideal environments for species such as dragonflies and frogs, which live in water environments early in life and move to woodland and grassland environments as adults. Similarly, birds of prey often breed in woodlands, but hunt in grasslands and wetlands.

Human requirements for a variety of natural resources enabled species diversity by providing a variety of habitats. This also met the needs of species that require several different environments to survive. In other words, the traditional production process and demand for natural resources created a combination of different habitats that resulted in high diversity of satoyama landscape species. The spatial characteristic of natural mosaics of satoyama environments is highly dependent on existing natural conditions. For example, topography and hydrology are important control factors in hilly landscapes and upland edge environments because the distribution of paddy fields and other croplands constructs a frame of environmental mosaics. On the other hand, environmental mosaics are mainly driven by human demand in relatively homogeneous landscapes such as uplands.

Environmental mosaics, although intended for natural resource harvest, are based on different types of nature depending on the region. Therefore, while sharing common characteristics, satoyama ecosystems in different regions consist of different combinations of species. That is, γ diversity of satoyama environmental variations can be perceived spatially (e.g. at the scale of Japan or East Asia). Because of γ diversity, the importance of biodiversity conservation can be justified in any region.

Conservation ecology is the study of relationships between beings and their environment, with the sociological goal of conserving biodiversity and maintaining ecological integrity. The nature of satoyama landscapes, which is a combination of paddy fields and satoyama landscapes, is a very important subject in conservation ecology today.

2.3 Citizen Conservation of Satoyama Landscapes

N. Kuramoto

2.3.1 Origins of the Nature Management Movement by Local Citizens

The nature conservation movement of the 1960s in Japan was focused on conserving natural areas near human settlement while simultaneously opposing development on important remote natural areas. An effective method of preserving local nature was through local residents learning to recognize the beauty and wonder of nature through various events and activities.

The conservation of satoyama landscapes required both the cessation of destructive development and adequate management. In the late 1980s, local movements began to appear that focused on satoyama landscape management.

The earliest satoyama landscape conservation activity of organizations found in Tokyo was the management of vegetation so as to preserve symbolic species of satoyama landscapes. Nature conservation activities were started by various organizations including The Organization for Preserving *Erythronium japonicum* in Shimizu Yama of Nerima City, Tokyo in 1976, The Organization for Preserving the District Flower: *Anemone flaccida* of Itabashi City, Tokyo in 1982, the Sengen Yama Nature Conservation Organization, which focused on preserving *Hemerocallis aurantiaca*, of Fuchu City in 1982, and The Organization for Preserving *Katakago* (the old Japanese name for *E. japonicum*) Forest in Machida in 1986. Activities that focused on satoyama landscape conservation included the West Tama nature forum by Kashinoki-yama Natural Park Lovers Association in 1986, and those of Mixed Woodland Volunteers in Sakuragaoka Park. The focus was enlarged from preserving specific species to preserving and managing the entire satoyama landscape environment.

However, satoyama landscapes in which development and nature-loving local citizens clash and satoyama landscapes where citizens are involved in management do not always coincide. Whether the satoyama landscape is located on private property or not, and whether or not the landowner has an understanding of conservation, can have much to do with public participation in landscape management. Public education can be very effective in enhancing participation in landscape management. Establishing trusting relationships between landowners and local citizens is also effective in preserving areas from development.

2.3.2 Satoyama Management by Citizens

There are many differences between satoyama and yatsuda landscapes in terms of management activities, participants' objectives, legal restrictions and local practices (Enari 2000; Table 2.1). In Takusi Enari's presentation of satoyama landscape task groups that worked with networks developed through the internet, a comparison of public activities in satoyama and yatsuda landscapes was made. The most specific difference between the two types of landscape is that there are few legal restrictions to public participation in woodland management (satoyama landscapes), but many for agricultural lands (yatsuda landscapes). By law, agricultural lands are not permitted to be managed or owned by individuals other than farmers. Legal cultivation by the general public can only be done in public allotment gardens or by volunteering to help existing farmers (Enari 2000).

The resulting products from woodland management activities are not wood products, as might be expected, but, at the moment, are only by-products or amenities. In most cases, these products do not guarantee the financial sustainability of management activities.

Allotment gardens that are set up under the Law of Allotment gardens are run by municipalities or agricultural corporations. They have high-quality gardens and equipment with which to manage the land. Volunteering to help farmers with agricultural activities has been popular in Yokohama City and Nerima City. Relationships are now being established between existing agricultural communities and local residents. This would provide locals with an understanding of agricultural activities and would also include economic contributions to agriculture through activities such as purchasing agricultural products, seeds, and seedlings (Enari 2000).

2.3.3 Satoyama Landscape Volunteers

The conservation and management of satoyama landscapes by local citizens is occurring in many parts of the country. There are 69 organizations recorded in

Groups Aiming for Regeneration and Conservation of Coppice Woodlands by Nakagawa (1996b).

Apart from classical agricultural land management, various forms of activities are present. This indicates that we are currently in a transition period waiting for a procedure to be established as a prototype (Kagawa 1999). It is more likely that variations of activities will persist, because every situation is site specific and depends on the natural condition, local participants, and the form of land ownership.

The activities can also be classified by relationships between landowners and municipalities. That is, classification depends on whether the land is public or private property, and also on the landowner's relationship with the municipality.

Examples of public property management in cooperation with municipality and citizens are the Sakuragaoka Park Coppice Woodland Volunteers and Sanaruko Lake Satoyama School. Sakuragaoka Park Coppice Woodland Volunteers is an example of citizens managing coppice woodlands on public lands while maintaining a close relationship with the municipality. As part of the metropolis park management, Tokyo Metropolitan Park Association supports the financing of this project.

There are numerous organizations that coordinate activities on public lands, but relationships with municipalities are often strained. Many organizations voluntarily organize nature watching and patrolling in parks and nature conservation areas.

Coppice woodlands found within private property are rapidly disappearing, mainly as a result of ownership changes through inheritance. In order to conserve private property coppice woodlands, some groups are lobbying for the reformation of the tax system or some other economic advantage for coppice woodland conservation (for example Symposium in Preserving Coppice Woodland in Tama held in 1997).

The Tamagawa Kizuna Woodland (Nakagawa 1996) is an example where good relationships are maintained between municipalities through citizen participation in coppice woodland management on private property. In Kanagawa Prefecture, the Kizuna Woodland Plantation Enterprise was involved in woodland management for 5 years, from 1987 to 1992. It was a plan that aimed to ally three groups: municipality, landowners, and local residents who want to use the coppice woodland. Local residents have access to the woodland for free, manage and maintain it in return, while appreciating the woodland nature. The Prefecture has supported the local residents financially. The residents' involvement in the management of the woodland started with mowing the undergrowth, which was thick with *Pleioblastus chino*. The activities have been carried out in a flexible manner while establishing relationships with a local community center, a nursery school, and an elementary school.

The Yokohama Woodland Forum is an example of networking activities with municipal connections. Yokohama City has been conducting the Yokohama Woodland Promoting Enterprise since 1994. The Yokohama Woodland Forum, which consists of 10 public organizations, undertakes networking activities. In 1997, 3000 citizens participated in plantation activities. They have been conducting voluntary management, which is only possible by local residents (Tsuchiya 1998).

Table 2.1. Woodlands and farmlands as fields of citizens' activities (Enari 2000)

Items	Woodlands	Farmland
Land as target	Green tracts of land in cities; coppice woodland near cities; artificial woodland in mountains Privated land→partly public land	Fields and non-cultivated land in and near city; paddy fields near city; fields in mountains. All is private land
Scale of field	Large-scale woodlands (0.1–100 ha)	<Field>Small section (10–1000 m^2) <Paddy field> Advance section (300–1000 m^2)
Details of activities	Part of forestation (plantation, mowing, middle felling etc.) Preparation of environment Processing and utilization of felled wood and by-products Nature watch, interchange →Part of fostering management and utilization of environment and by-products	Help-farming; leased field Processing, utilization and food of harvests Nature watch, interchange All of agriculture and food (cultivation–harvest–food)
Forms of activities	Leaders are essential. Power machines and specific tools are mainly used Space and tools are needed for processing →Group cooperation work	<Field> Possible without leader Possible with gardening tools Sometimes completion in-house →Mainly private activity <Paddy field> Leader is needed Exclusive farm machines are used Adjustment and processing facilities are needed →Group cooperation work
Frequency and period of activities	1–30 days/year Long unit longer than several years	<Field> 10–100 days/year Unit is a season <Paddy field> 5–100 days/year Unit is a year
Aims of participation	Labor in nature Interchange of communities Nature study, nature experiences Field work Woodworking, handcrafts Education of children/restoration of environment and landscape	Labor in nature Interchange of communities Nature study, nature experience Fresh vegetables Production and supply of food Education of children/restoration of environment and landscape
Outcome	By-products: processed woods, firewood and charcoal, mushrooms Main are non-essential items for life →Leisure oriented	Farm products: vegetables, grain etc. Processed products: mainly food →Part of food and daily necessities →Living oriented

Items	Woodlands	Farmland
Necessary skills and quality of work	A small kind and large quantity of physical labor, dealing with large massive matter Power tools (chain-saw, mower etc.); long time before result (one to several years) →Dynamic collaboration of hard work →Opportunity of participation is rare; conditions are severe	Various cultivation techniques, dealing with much small and light matter Tools for handy work (hoes and sickles) Short time before results (several days to several months) →Private work accumulating fine processes <Field> Intensive joint work in a limited period <Paddy field> Intensive joint work →Opportunity of participation is frequent Conditions are easy
Expectation for citizens by owner	Proper management of environment and vegetation; conservation of scenery Protection against disaster No expectation of forestry →Maintenance and improvement of existence value	Maintenance and succession of productivity as production and harvest site Expectation of farmland; assessment in fixed property tax →Maintenance and improvement of working value
Backgrounds of industry and economy	No forestry managers Decline in industrial value Flats are subject to development →Expensive tax →Left as not industrial use No manager	Many forefront farm managers Valuable place for production and watched as property →Expensive tax →Existence of many professionals as bases of management
Restriction of citizens' act	Establishment of right (ownership and lease) is possible Promotion of citizens' participation is increasing in administrative measures →No joining control except specific area	Impossibility of title establishment (ownership and lease) Farming is impossible in principle
Relationship with natural ecosystem	Appropriate management of woodland environment has priority Activation of forest and trees Multilayer of forest Diversification of woodland floor animals and plants	Artificial crop production has priority Exclusion of weeds and harmful insects →Crops have priority/ maintenance of productive environment Increase of animals and plants adapted to circumference and fallow land Influences of agricultural chemicals and chemical fertilizers

2.3.4 Goals of Public Conservation of Satoyama Landscapes

The goals of satoyama landscape conservation depend on whether the focus is on the satoyama landscape itself, on the local residents, or on the landowners.

The satoyama landscape goal is conservation of species diversity, an aesthetically pleasing landscape, and the traditional culture and techniques of satoyama landscapes. The local resident goal is to provide new outdoor recreational opportunities, enhanced quality of life, construction of networks, and establishment of local autonomy. The landowner goal is to gain empathetic understanding of local residents in maintaining the landscape.

Satoyama landscapes have gained recognition in the Red Data Book of plant species for the conservation of species diversity (NACS-J, WWFJ 1989). This book includes endangered species that inhabit satoyama environments (Kameyama et al. 1994). Satoyama environments are also considered to function as refuges for species such as *E. japonicum*, which has survived since the glacial period (Moriyama 1988).

The overall goal is to preserve the traditional productive satoyama landscapes for future generations. Many coppice woodlands are now over 40 years old, which is much older than previous coppice woodlands. Satoyama landscape management can revive the use of traditional satoyama landscapes as a part of daily life.

Satoyama landscapes were originally agricultural woodlands and agricultural techniques existed relating specifically to their management. Cultural practices such as *yamahajime* (the opening of woodlands regulated by local customs) and *imibi* (woodland access restricted by local customs) also exist. These cultural practices are disappearing and it is important to preserve these practices for future generations. These techniques and cultural practices are regional. Even in the same Tokyo Metropolis, practices found in the Sayama Hills and the Tama Hills are different (Kuramoto et al. 2001). Merely inheriting techniques and culture from a specific place does not help when the rest of the local practices are disappearing. These connections with satoyama landscapes and traditional ways are related to the current goals of conservation.

Satoyama conservation management has great recreational potential, as is apparent by the number of citizen volunteers. In other parks, users take part in planned recreational activities but, with satoyama management, the public enjoys a new form of recreation that involves them as they construct their own recreational open space.

Satisfaction in life comes from the fact that satoyama requires human intervention. This human intervention must, however, be in balance with natural successional forces. From a citizen's perspective, citizens are needed for the existence of the woodland. It is important to be needed. Everyone can solve their problems by becoming volunteers (Takahashi and Hatano 1990). In addition, people seek to understand the meaning of their existence, especially the need to help someone or something else (Kamiya 1980). Satisfaction in life can be attained by becoming aware of the necessity of one's existence. There are two types of satisfaction in life: that gained by expressing oneself; and that gained by being needed. Satoyama

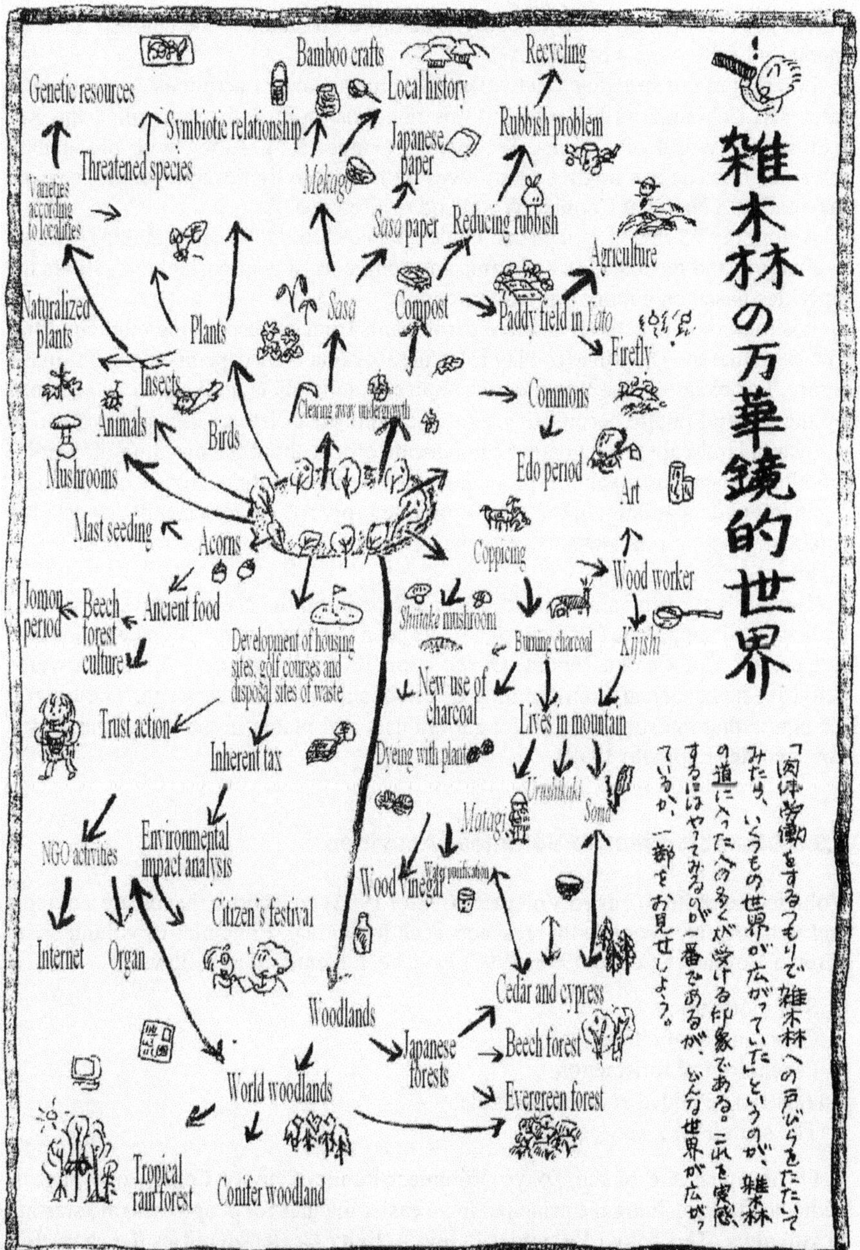

Fig. 2. 7. The kaleidoscopic world of the coppice woodlands (Uchinuma 1997)

volunteers fall into the latter category. The more satisfaction one attains, the more mentally healthy one's life is.

In terms of constructing a network, satoyama volunteer activities open up more interests. Uchinuma (1997) showed this relationship in a diagram called the Kaleidoscopic World of Coppice Woodlands (Figure 2.7). Along with broadening interests, citizens can involve themselves with more activities and people, or even take part in a National Coppice Woodland Conference.

Autonomy of citizen groups can be developed through decision making, action, evaluation, and feedback. In satoyama environments, it is important to evaluate the activities based on a monitoring of outcomes.

It is important to actively involve participants from corresponding municipalities to ensure that they continue to play roles in satoyama landscape management in the future. For example, the thinning of coppice woodlands is unlikely to be accepted by the general public because it means the removal of trees. Citizens involved in satoyama landscape management can educate others about the meaning of coppice woodland management and its procedures. Thus, participating citizens can promote a general understanding. In the case of private property, citizens can be expected to establish trusting relationships, and even economically beneficial relationships, as mentioned previously.

The goals for satoyama landscapes can depend on the stage of development as well as local situations. For example, it has been reported in the Coppice Woodland Organization of Ushiku Natural Observation Woodland that the attitude towards activities has changed from the time of establishment of 6 years ago. The biggest change is that the subject is no longer animals and plants in the woodland, but is now people (Enomoto 1999).

2.3.5 Characteristics of Volunteer Activities

Volunteer activities (Tokyo Volunteer Center 1996) are actions that create a diverse and rich society, a society that respects each individual. Principles of volunteerism (Tokyo Volunteer Center 1996, 1997) have been identified as follows:

1. Self motivation;
2. Cooperation of citizens; and
3. Psychological satisfaction.

To this can be added another principle:

4. The search for a new society.

On the web site of the Tokyo Volunteer/Public Activity Center, one page is dedicated to explaining the principal in an easier manner for people who are starting to volunteer. The Tokyo Volunteer/Public Activity Center identifies the characteristics of volunteers today as:

1. From self-sacrifice to self-growth;
2. From one-way relationships to inter-relationships;
3. From hierarchical relationships to equal citizen relationships;

Fig. 2. 8. Planning by a satoyama volunteer (Volunteers for Conservation of Coppice in Sakuragaoka Park; VCCS)

4. From a special event to daily activity; and
5. From standardized society to a society with a multivalue system.

One prototype of a volunteer activity is as follows. The participants consider themselves as equals and, as a group, aim to maintain the natural integrity of the conservation area. Among the participants, there are likely to be several talented individuals who can lead specific activities, such as knowing how and who can take the lead in thinning, or the editing and publishing of newsletters and so on. Scheduling of a day's activities can be led by the person whose turn it is to be in charge. Plans are made by a committee a priori. Decisions are made collectively by all responsible participants, including volunteer participants and committee members (Figure 2.8). Plans are self-evaluated and results are fed back to the original plan (Figure 2.9). The leaders also make activities public by providing opportunities for experience as well as nature watching, to allow others to share the joy of being involved with satoyama nature, and also to gain public understanding. Volunteer coordinators, who find the connections and coordinate the relationships, are very important in such activities.

Fig. 2. 9. VCCS volunteers regularly measure the length of sprouts from stumps

2.3.6 Coordinators of Volunteers

It is important to have coordinators who can link people with satoyama landscapes. The following outlines a relationship that a coordinator might have to deal with, and what is needed for a coordinator to run activities (Kuramoto and Imoto 1998, Fugure 2.10).

There are five relationships that a coordinator must deal with: public and nature; public and public; public and region; public and network; and public and municipality.

1. Public and Nature: Satoyama landscapes are a consequence of agricultural activities. The resulting ecosystem will be completely changed if human intervention ceases and natural succession progresses. Today, agricultural activities have changed and satoyama landscapes are being neglected. Deliberate management in place of agricultural activity is necessary in order to maintain the unique satoyama environment. The coordinator plays the role of linking citizens with satoyama landscapes, which, in turn, need the involvement of humans to survive (Figure 2.11).
2. Public and Public: Public involvement in satoyama landscapes tends to be by large groups with common interests, rather than through traditional family involvement. It is because of the need for information exchange with other inter-

Fig. 2. 10. The Networking Program by satoyama volunteers (VCCS). Volunteers hand out oak acorns to children for planting in order to grow new trees

ested citizens and the size of large-scale projects that it is necessary to have a coordinator to make connections. Functional organizations can make good use of every individual, and each individual has an important role to play.

3. Public and Region: In terms of understanding and adhering to traditional local culture and techniques, a leader must be from the local area. Recording the human–nature relationships of an area is very important. For a long-term activity, it is invaluable to involve local individuals.
4. Public and Network: Publicly managed satoyama landscapes extend across Japan. Many activities have occurred through the assistance of networking and communication. For example, each organization can gain new procedures beyond its experience, and can locate itself in a social context. Local activities are not bureaucratic, nor are they a hierarchical system, but, rather, they are a system that can be established according to local requirements. Everyone can contribute by making the best use of their strengths. All together, the opportunity of attaining new interests in other disciplines can be provided. Everyone can take part in the network based on their experience. Coordinators can make the links between citizens and networks depending on their interests and aptitudes.
5. Citizen and Municipality: Especially when dealing with public property, it is often inevitable to have to deal with municipalities for satoyama activities in Japan. It is important to deal with them wisely and have a good understanding of their budgets, plans, decision-making procedures, and personnel administration systems.

Fig. 2. 11. The Sharing Program by VCCS volunteers provides an opportunity for children to cut wood

A comparison can be made between coordinators from municipalities and private consultants because their functions are more-or-less the same (Komazawa and Kuramoto 2001). This suggests that it is possible to establish professional coordinators.

2.3.7 Characteristics of Coordinators

Coordinators play very important roles for citizens involved with satoyama landscapes. The following are suggested attitudes needed for a coordinator.

In order to give breadth and depth to projects, an individual's strengths and abilities should be considered and respected. Careful consideration is necessary in order to have participants work where they are best suited. For example, someone who is not physically strong enough to thin trees may be able to chop the thinned trees. Coordination should be flexible enough so that people can participate to the full extent of their abilities.

It is important to be able to maintain one's composure when coordinating activities and participants. Projects do not always work out as the coordinator intended. It is important for coordinators to recognize that it may be necessary to seek help and advice from others.

It is necessary to not only take care of the matters within the organization, but also to create information networks among corresponding individuals and other organizations. The network will support the activities.

Coordinators can be personnel paid by the municipality, an instructor from the outside, or a consultant. They can also be unpaid volunteer citizens. The responsibilities of volunteer coordinators are somewhat heavier than for other volunteers. There is a demand among coordinators that the position be a paid position from the beginning.

2.4 Environmental Policy and Satoyama Landscapes

Y. Kitamura

2.4.1 Satoyama Landscapes and Previous Environmental Policy

2.4.1.1 Environmental Policy in Japan

There are two components of the Japanese natural environmental law system: The Natural Environmental Conservation Law and the Natural Parks Law. The former seeks to conserve high-quality nature through zoning that restricts activities. However, only five wilderness areas have been designated in Japan, with the most recent designation being in 1985. In addition, only 10 nature conservation areas exist, including the Shirakami Mountains established in 1996, the first such designation in 13 years.

Under the Natural Parks Law, national parks, quasi-national parks, and local natural park systems can be designated. Activity restrictions are set up within each area. However, unlike The Natural Environmental Conservation Law, the Natural Parks Law does not focus on ecological integrity, but on landscape aesthetics. This is clear from the description of the law as the protection of prominent landscapes and promotion of their public use (Article 1, the Natural Parks Law).

The goal of the Natural Environmental Conservation Law is to protect remote areas. It is often said that the law respects lineage because areas tends to be selected on the basis of uniqueness, rareness, and academic importance. On the other hand, the Natural Parks Law aims to protect only one aspect: high scenic value. The attitude of only focusing on the positive or aesthetic part of nature is understandable considering the context. The Natural Parks Law was developed at a time when Japan was losing its scenic natural environment. There is an obvious limitation to this conservation law in that, under its terms, satoyama landscapes generally cannot be subject to conservation restrictions.

Unlike the areas designated under these two laws, satoyama landscapes are often local nature taken for granted. Satoyama landscapes not only have physical and ecological functions, but also sociological functions within the local community.

This was beyond the scope of previous nature policies. A policy that genuinely dealt with all the functions of satoyama environments has been a long time coming.

2.4.1.2 Municipal Environmental Policy

From the 1970s onwards, some local municipalities began to prepare environmental management plans. It is noteworthy that the impetus for these plans was not a desire to recover from disaster, but a concept of schematic management of the local environment. However, the actual substance of these plans usually came down to dealing with the seven typical pollutants plus the natural environment, which was the original mandate of the Environmental Agency at the time. Municipalities were not concerned about the public benefit of satoyama landscapes at this time. When satoyama landscapes happened to be included in site planning, it was never discussed from the perspective of agriculture, but focused only on the environmental. There was a big problem with the execution because no standard legal procedure existed.

2.4.2 Satoyama Landscapes and Recent Environmental Policy

2.4.2.1 Environmental Law and Environmental Planning

As discussed earlier, satoyama landscapes that were neither in urbanized nor natural areas were, at one time, not considered a subject worthy of natural policy. However, in the past 20 years there has been increasing public interest in the many functions of secondary nature and a demand for an improved quality of life. These concerns have manifested themselves in policies that promote improved human–nature relationships. The term *coexistence between nature and humans* reflects this. The Basic Environmental Law, enacted in 1993, which combined the basic policies of the Basic Law for Environmental Pollution Control and the Natural Environmental Conservation Law, indicates the concept of coexistence, which was not included in the past two laws.

In addition to the previous policies that dealt with pollution and nature protection in the policy enactment guide, the Basic Environmental Law added a new policy objective, in Article 14, to address satoyama landscapes, including satoyama landscape conservation. This article sought to ensure that diverse natural environments, such as woodlands, agricultural lands, and wetlands, are systematically conserved in their local natural and social context, while conserving ecological diversity, preserving wildlife, and conserving other species diversity to make sure that humans and nature get involved with each other.

Sustainable development is an essential concept of the Basic Environment Law, which should also be carefully considered when conserving satoyama landscapes.

The policy substance is discussed in more detail within the Basic Environmental Plan based on Article 15 of the Basic Environmental Law (Section 3, Chapter 2, Paragraph 1.6). The terms *satoyama woodland* and *yatsuda* are referred to in these sections, and public involvement with nature is also particularly emphasized. The previous plan, before it was revised in December, 2000, discussed in detail the function and direction of the policy. However, in the new plan, the function and direction is reduced to the formation and maintenance of the secondary natural environment. The term *satoyama* no longer appears in the title, but the basic policy remains the same (The Basic Environmental Law, Article 15, clause 3).

Thus, it can be seen that satoyama landscapes have been a concern of the Japanese government, at least superficially. The Basic Environmental Plan is a concept statement of the National Policy that must be operationalized by appropriate departments.

2.4.2.2 Law of Food, Agriculture, and Rural Community

The Basic Law of Food, Agriculture, and Rural Community, enacted in 1999, took over from the Basic Agriculture Law and, to an extent, regulates satoyama landscape policy direction. With regard to the agricultural communities within satoyama landscapes, The Basic Law of Food, Agriculture, and Rural Community states that the multifunction of the national land, water, natural environment, and aesthetic landscape conservation, and cultural heritage that local agricultural activities provide have to be best and appropriately displayed throughout the future (3rd Article). Consideration of environmental values can also be seen in the modification of the Riparian Act. The political interests in the agricultural field, in a broader sense, are increasing, which was not the case during the time of the Basic Agriculture Law.

2.4.2.3 Autonomous Environmental Planning

In the 1990s, basic environmental plans were enacted in many local municipalities under the Basic Environmental By-laws. The substance of the plans varies from municipality to municipality, under the guidance of the Local Environmental Planning Division of the former Environment Agency. In light of the trend toward environmental planning by local municipalities since the 1970s, the Agency produced a guideline as a prototype of local environmental planning that acted as a milestone in the new Basic Environmental Law era. It is presumed to be referred to by local autonomy when applying the concept of the Basic Environmental Law. It has had more than a small influence on local policy.

In the guidelines, a satoyama landscape has not been specifically defined. It has been recognized spatially as an area that is mostly occupied with woodland and agricultural land, or an area that is located in the middle of naturalness: an area that is mostly occupied by woodland and agricultural land with a decreasing and aging population that will become increasingly unable to maintain it. The guidelines also state that it is important to:

'... go back to the basics and reexamine how to appropriately manage and use the woodland and agricultural land, while improving the health of local economies and minimizing impacts on the natural environment.'

Because these landscapes can be adjacent to urban areas, the guidelines identify one of the main problems to be that of minimizing environmental impacts due to industry and population inflows to the area. The guidelines also say that it is important to plan social capital arrangements for environmental conservation and land use modifications, as well as to prevent environmental deterioration prior to development. In order to do so, there is a need to promote the creation of a community of public participants to live in harmony with the natural environment.

2.4.3 Satoyama Conservation in Municipal Planning

The former Basic Environment Plan, which was enacted in 1994, had certain policies under the name of *Satoyama Landscape Areas*. So, how is the nature of satoyama landscapes treated in the basic environmental plans of municipalities? The following are some examples.

The Miyazaki Prefecture Basic Environmental Plan was enacted in 1997. It designated the entire prefecture into categories of plains, satoyama landscapes, mountainous lands, and coastal and riverside lands. For satoyama landscapes, the present conditions, problems, and matters for consideration have been described. Among them is the fact that the public benefit of agricultural lands is being adversely affected by land neglect as young people migrate out and the elderly population increases, and as a result of decreasing secondary woodlands due to tourist-related developments.

Other matters of consideration include: the ecosystems of the woodlands and agricultural lands; adequate land use strategies for tourist-related developments; adequate management and maintenance of satoyama woodlands; and yatsuda environments that require community involvement.

The Gunma Prefecture Basic Environment Plan enacted in 1997 similarly classifies land as mountainous areas, satoyama areas, flat areas, and wetland areas. In this plan, satoyama areas are intended to be landscapes where native plants can be preserved, and also areas where environmentally friendly industries, such as environmentally friendly agriculture, can be encouraged in order to allow humans and nature to coexist.

The Aichi Prefecture Basic Environment Plan was enacted in 1997. In this plan, measures for integrated satoyama conservation include adequate land use strategies, promotion of public involvement with satoyama landscape management systems, and so forth.

As for cities, the Akashi City Basic Environment Plan was enacted in 2000. The Akashi City Basic Environment Plan states that positive conservation activities will be undertaken for woodland conservation in urban and surrounding environments, including satoyama landscapes. The 21' Iida Environment Plan (Iida City, Nagano Prefecture 1996) includes detailed proposals, such as biodiversity conservation designation with bylaws concerning satoyama landscape conservation.

2.4.4 Satoyama Landscape Conservation Strategy

Thus, even in environmental policy, recognition of the biological and social meanings of satoyama landscapes has been established. The previous examples lack real substance, but this is inevitable considering that the Basic Environmental Law, upon which they are based, is very theoretical.

Just as the Basic Environmental Plan follows the Basic Environmental Law, the Basic Environmental Plans of municipalities follows Basic Environmental By-laws. For enactment to occur, the Basic Environmental Plan needs to be approved by the entire department. In most cases, the environmental sector is the main office the by-law needs to be approved by, but the urban planning and agricultural sectors should agree to it as well.

There are several ways that satoyama landscape conservation plans are executed. One is by regulation. The purpose behind this is to stop chaotic development by legally controlling land use. The existing law can be adapted, or a new law can be established. Another way is through economic restriction. The purpose of this is to promote satoyama landscape functions by financially supporting farmers and non-profit organization (NPO) activists, or by giving incentives, such as tax reductions. Developers can also be legally charged for the social expenses incurred as a direct result of their development. Regardless of these issues, corresponding departments need to cooperate to move on to the next stage of planning in order to realize these plans.

2.4.5 Satoyama Conservation in Municipalities

The relationship between local and prefecture municipalities is an important consideration in satoyama landscape conservation. As mentioned before, each basic environmental plan of both prefecture and local municipalities states existing conditions and problems from their own perspective. Satoyama landscapes that are subject to a city plan can also be subject to a prefecture plan.

Generally, in basic environmental plans, there is no adjustment between overlapping municipalities for each subject. As will be discussed in Chapter 7, it is often the city that takes action in satoyama landscape conservation. This is not to say that satoyama landscape policy belongs only to local municipalities, but that the relationship and roles of the overlapping municipalities need to be examined in the future.

3 Satoyama Landscape Transition

3.1 Transition of Satoyama Landscapes in Japan

A. Tsunekawa

3.1.1 National Scale Transition of Land Use

A British botanist by the name of Robert Fortune visited the Yokohama area of Japan in 1860, at the end of the Edo era. His writings described the traditional satoyama landscape (Fortune 1863):

> I often walked into farm communities. The houses were built on cultivated lands at the foot of the mountains. In behind, there were woodlands and in front there were paddy fields. Roofs were thatched, resembling those of temples but not as fine and solid. It was a typical view of countryside that there were, without any exception, irises growing on top of the roofs (Yedo and Beijing).

Whereas in 1860 Japan was a rural, agricultural country, less than a century later farming as an occupation had dropped to 45.2% (1950) and then to 6.4% (1990) of the population. Forestry as an occupation had dropped similarly to 1.2% and then to 0.2%, respectively. Clearly, Japanese agriculture has a much smaller role than it did a century ago.

What has been the effect of this change on rural land use? A group led by Professor Yukio Himiyama of Hokkaido University of Education is developing a national scale database to address this question (Himiyama 1995). Data from four time periods were analyzed: about 1850, 1900, 1950, and 1985. A land-use map was developed at a scale of 1 : 50 000. Land use was identified for each 2 km^2 cell, then the cell numbers were counted to allow transformation into the total area for each land-use category. Table 3.1 shows the land use of the four time periods.

Analysis indicated that forests and woodlands covered 66.6% of the total area in 1985. This was 2.0% less than that in 1950 and 1.1% more than that in 1900. This indicates that forests and woodlands have not changed greatly in area, on a national scale, in 100 years. This is a surprising result, considering the huge social and economic changes during this time.

However, detailed analysis revealed that *hinoki* or Japanese cypress (*Chamaecyparis obtusa*) and *sugi* or Japanese ceder (*Cryptomeria japonica*) plantations in

Table 3.1. Changes in national land use since 1850 (Source: Himiyama 1995)

	ca. 1850		ca. 1900		ca. 1950		ca. 1985	
	×10km^2	%	×10km^2	%	×10km^2	%	×10km^2	%
Urban land use	1064	2.86	1548	4.16	2269	6.10	3767	10.11
Built-up area	612	1.65	645	1.73	975	2.62	2032	5.45
Road	452	1.22	862	2.32	1202	3.23	1645	4.41
Rail road	0	0.00	41	0.11	92	0.25	90	0.24
Agricultural land use	5345	14.38	6222	16.72	6303	16.94	6417	17.21
Rice field	3592	9.66	3462	9.30	3623	9.74	3573	9.59
Farm	1752	4.71	2323	6.24	2224	5.98	2179	5.85
Mulberry field	0	0.00	343	0.92	352	0.95	145	0.39
Tea plantation	0	0.00	17	0.04	14	0.04	50	0.13
Orchard	0	0.00	40	0.11	76	0.20	454	1.22
Other tree field	1	0.00	38	0.10	13	0.04	16	0.04
Forests and woodlands	25 497	68.59	24 348	65.44	24 885	66.88	24 818	66.58
Broad-leaved forests and woodlands	11 176	30.07	9856	26.49	9484	25.49	5314	14.25
Coniferous woodlands	4572	12.30	4410	11.85	4377	11.76	3854	10.34
Mixed woodland	9426	25.36	9773	26.27	10 757	28.91	15 138	40.61
Bamboo woodland etc.	323	0.87	309	0.83	268	0.72	512	1.37
Others	5265	14.16	5090	13.68	3751	10.08	2274	6.10
Waste land	4401	11.84	4180	11.23	2917	7.84	1372	3.68
Wet land	161	0.43	152	0.41	101	0.27	73	0.20
Paddy field	702	1.89	722	1.94	724	1.95	760	2.04
Golf course	0	0.00	0	0.00	0	0.00	68	0.18
Others	2	0.00	36	0.10	8	0.02	0	0.00
Total	37 171	100.00	37 207	100.00	37 207	100.00	37 275	100.00

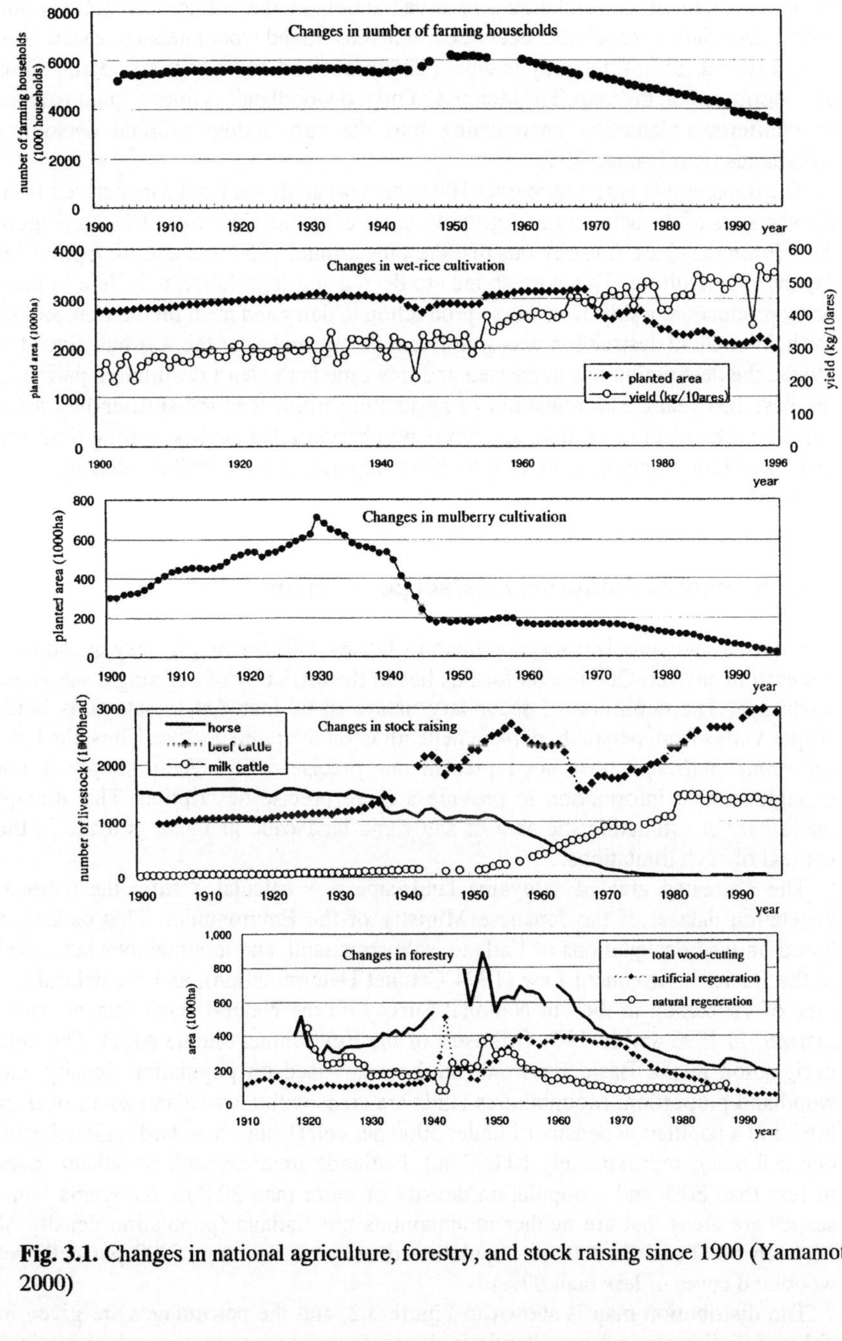

Fig. 3.1. Changes in national agriculture, forestry, and stock raising since 1900 (Yamamoto 2000)

Chugoku, Shikoku, and Kyushu had offset woodland destruction due to agricultural land development in Hokkaido and urbanization in the Kanto area (Himiyama 1995). Deciduous woodlands decreased, whereas mixed woodlands increased. The mixed woodland in this study is where deciduous and coniferous woodland signs are intermixed on the map. The increase of mixed woodlands is due to small patches of coniferous plantation encroaching into the surrounding original deciduous woodlands (Himiyama 1995).

The transition in agriculture over 100 years was analyzed from data derived from the statistics of the Ministry of Agriculture, Forestry and Fisheries of Japan (Figure 3.1). From the chart, it is obvious that the time around 1950 was a turning point for Japanese agriculture. This corresponds to decreasing farm labor, a decline in mulberry production, a shift from horse production to dairy and meat production, and so forth. Woodland destruction was significant during and after the war but, since the 1980s, the destruction has decreased and this time is the least destructive period in the past 100 years. The transition of agriculture influenced the surrounding satoyama landscape. For example, less wood was harvested at the best cutting time and there was little thinning, both of which are imperative for woodland density management.

3.1.2 Amount of Satoyama Landscape in Japan

How much satoyama landscape is there in Japan? This seemingly easy question is not easy to answer. One reason for this lies in the difficulty of defining a satoyama landscape. The definition of secondary nature to be included in satoyama landscapes varies from person to person, depending on their perspective. Thus, the term *satoyama landscape* does not represent one precise entity. There simply is not enough reliable information to provide a more precise description. The attempt made here at estimating the area of satoyama landscape in Japan is made in the context of such limitations.

The estimated area of satoyama landscape was calculated from the national vegetation dataset of the Japanese Ministry of the Environment. This dataset is based on land designations of flatland, satoyama land, and mountainous land used in the Basic Environment Law (1994 Cabinet Determination), and the natural degree of vegetation in the 4th National Survey on the Natural Environment summarized in 1993 and in 1994 (Ministry of the Environment home page). The land designation in the Basic Environment Law is based on population density and woodland proportion. Mountainous lands are areas with a woodland cover of over 80% and a population density of under 5000 per cell (from a standard regional grid, one cell being approximately 10 000 ha). Flatlands are areas with woodland cover of less than 80% and a population density of more than 30 000. Satoyama landscapes are areas that are neither mountainous nor flatland (population density of between 5000 and 30 000, or population density of less than 5000 per cell, yet woodland cover of less than 80%).

The distribution map is shown in Figure 3.2, and the percentages are given in Table 3.2. Forests and woodlands in Japan (natural, secondary, and plantation)

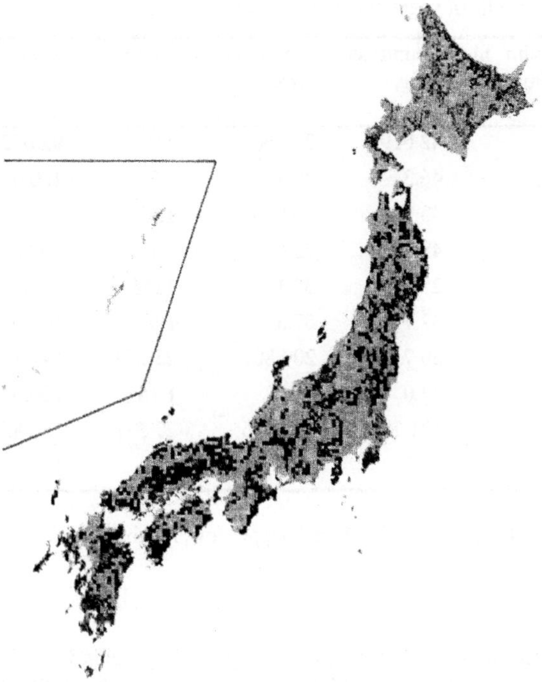

Fig. 3.2. Distribution of secondary woodlands and plantation areas (black dots) in the countryside.Changes in national agriculture, forestry, and stock raising since 1900 (Yamamoto 2000)

occupy 247 717 cells (63.7% of national land), with 66 394 cells occupied by natural forest (17.1%), 89 251 cells of secondary woodland (23.0%), and 92 072 cells of plantation woodland (23.7%). Secondary and plantation woodlands in mountainous land, satoyama landscapes, and flatland were as follows: 66 965 cells (17.2%) of mountainous land; 85 995 cells (22.1%) of satoyama; and 25 336 cells (6.5%) of flatland. Secondary and plantation woodlands in satoyama land is approximately 86 000 km^2 in total. Half of this is plantation woodland.

On the other hand, in their Satoyama Area Development Investigation, the Forest Agency of Japan (1978) defined satoyama woodlands as *young deciduous woodland*, which may be:

1. Private woodlands or national forest of the classification partly wooded or woodlands in common use;
2. Artificial woodlands of over 75% deciduous, which was used for charcoal and forestry production but not for producing building material;
3. Younger than 50 years of age; and
4. More than 0.1 ha in area.

Table 3.2. Number of grid cells of vegetation type in each region

	Areas with high human impact	Countryside areas	Mountainous areas	Others	Total
Plantation	11 339	42 619	36 960	1154	92 072
	12.3	46.3	40.1	1.3	100.0
Secondary woodland	13 997	43 376	30 005	1873	89 251
	15.7	48.6	33.6	2.1	100.0
Natural forest	1938	25 039	38 151	1266	66 394
	2.9	37.7	57.5	1.9	100.0
Others	36 241	60 753	20 880	23 278	141 152
	25.7	43.0	14.8	16.5	100.0
Total	63 515	171 787	125 996	27 571	388 869
	16.3	44.2	32.4	7.1	100.0

Each grid cell corresponds to approximately 1 km^2. Upper row, number of grid-cells; lower row, percentage.

Following this definition, so-called coppice and mixed woodlands come under the definition of satoyama landscapes, which cover approximately 60 000 km^2, or slightly over 70% of the public natural forest and woodland area.

Two estimates of the area of satoyama landscapes in Japan have been calculated. The Japanese national vegetation dataset defined satoyama landscapes as secondary and plantation woodlands in populated countryside (i.e., a population density of 5000 to 30 000 or population density of less than 5000 with less than 80% woodland). Although grassland is also included in the definition of satoyama landscapes, because the grasslands are significantly smaller than the woodlands and because secondary grassland in the national vegetation dataset also includes golf courses, secondary grassland was not included in the estimation made here. In contrast, the Forest Agency of Japan defined a satoyama landscape as a private woodland or national forest of common use planted with the aim of charcoal production. Thus, estimations of the area of satoyama landscapes depend on: whether the satoyama landscape is considered to include national forest that is not of common use (former) or that which is in common use (latter); forests based on ecological systems (former), or on human systems (latter); and whether population density is taken into consideration (former) or not (latter). However, regardless of which variables are considered, the results are similar. From this exercise, it can be concluded that satoyama landscapes constitute approximately 20% of national land, approximately 60 000–90 000 km^2, setting aside differences in definitions and input data.

3.1.3 Regional Transition of Satoyama Landscapes

As was discussed earlier, there are basically two types of satoyama landscape:

1. Areas where forest and woodland distribution is largely determined by land formation, with forests and woodlands remaining primarily on slopes; and
2. Areas where there is a mixture of woodlands and grasslands on flatlands, typically found in uplands and terraces.

Socioeconomic factors, such as distance from the village, are often the determining factor of woodland distribution (Figure 2.4). Lowland areas are typically not wooded because of favorable hydrologic conditions in the lowlands, which encourage cultivation.

Despite the differences, both types of satoyama landscape are determined by natural landforms that have undergone significant change since the early modern age. In the following section, the transformation of the Tama Hills is described as an example of the former type of satoyama landscape and the Musashino Upland is described as an example of the latter type of satoyama landscape.

3.1.3.1 Transformation of Satoyama Landscapes in Hills

Figure 3.3 shows the land use of the Tsurumi River basin at intervals from 1881 to 1995. Land use in the early Meiji period (1881) was characterized by rice paddy fields in valley bottoms and lowlands, dry fields in uplands, and deciduous woodlands on steep slopes. Many of the villages are on the boundary of uplands and valley bottoms, and lowlands (Yokota 1973).

In 1908, the Yokohama Line railway was opened between East Kanagawa and Hachioji and, in 1955, the Denentoshi Line railway was opened between Mizonokuchi and Nagatsuda. The 1967 map shows the change in land use, mainly around the stations, such as Saginuma and Ichigao. The main trend was the transformation of both dry upland fields and lowland paddy fields into residential areas.

Then, in 1974, a land readjustment project was conducted by the Japanese Housing and Urban Development Corporation and the development of Kohoku New Town began. The characteristic landform of the hills was flattened and turned into a new suburb with wide roads and medium-rise residential apartments.

3.1.3.2 Transformation of Satoyama Landscapes in Uplands

The Musashino Uplands contained common woodlands and fields where local residents gathered charcoal for fuel, fodder and grass for fertilizer, *Miscanthus sinensis* and *Miscanthus sacchariflorus* grass for roof thatching and wild plants for food (Kimura and Ito 1960). In the Edo period, new agricultural fields were developed. During the Kyoho period (1716–1736) in particular, the government encouraged such development and many new agricultural fields, called *Musashino Shinden*, were created. In the new agricultural fields, each farmer was allotted flatland woodland of a certain area instead of sharing common woodlands (Yajima 1955). *Quercus acutissima* and *Quercus serrata* woodlands were typical in these flatland woodlands. These were cut down in 15–25 year cycles to make charcoal for

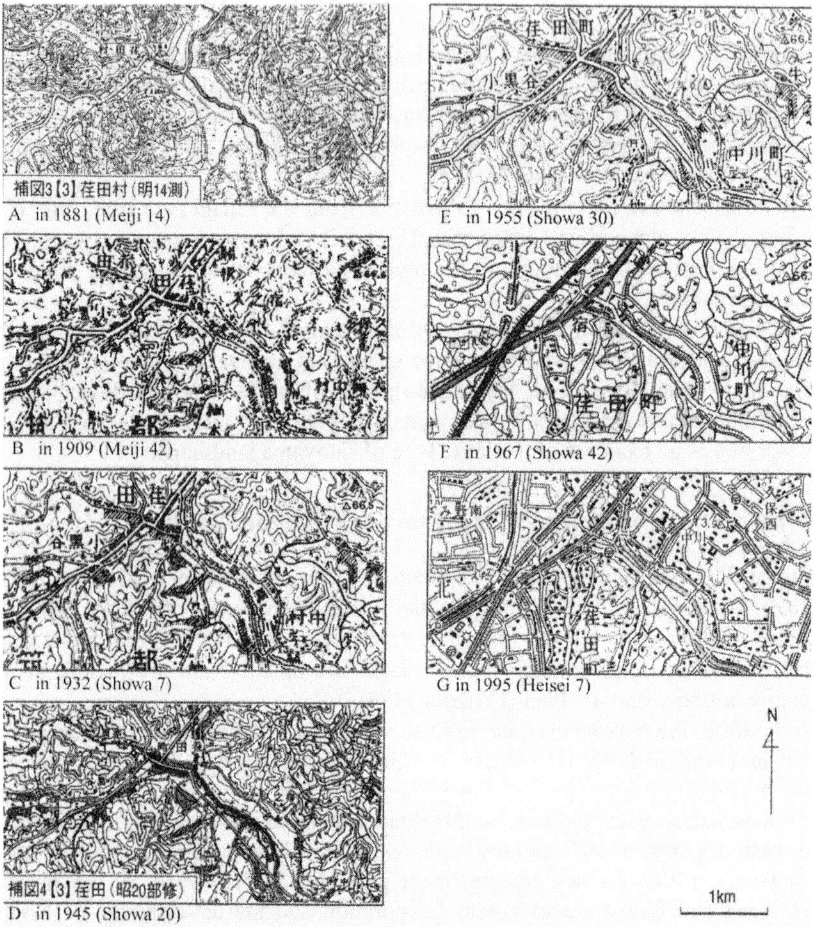

Fig. 3.3. Land use changes in the South Tama Hills, around Eda Village (Source: Japan Atlas System 1972)

fuel. Understory shrubs and grasses were mowed, and fallen leaves were gathered and used for fertilizer (Inui 1982). This satoyama landscape system remained after the new agricultural field development period until the Meiji period.

The flatland woodland in and around the Musashino Upland of 1880 (Figure 3.4) reveals that there were larger portions in the Sayama Hills and the Musashino Uplands, but not in the alluvial lowlands of the Tamagawa River and the Arakawa River, which were primarily paddy fields. The band of flatland woodland to the north-east of Sayama Lake (Figure 3.4) is the then new Santome Agricultural Field.

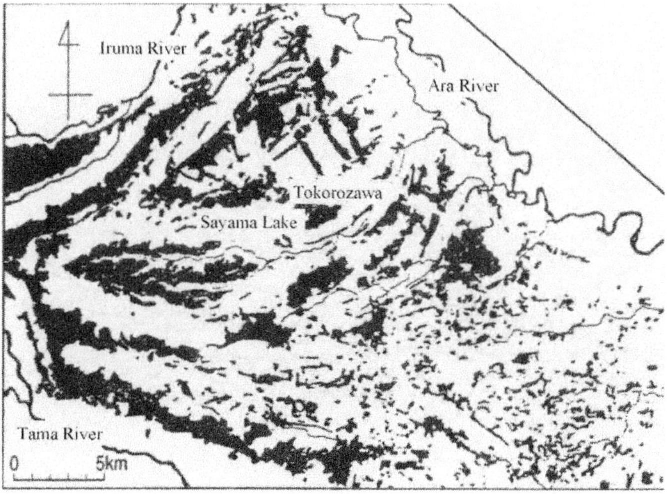

Fig. 3.4. Distribution of flatland woodlands on the Musashino Upland in the 1880s (After Inui 1982)

In creating the new agricultural field village, the land was divided into strips with flatland woodlands placed in the most remote locations.

During the Taisho period, rapid urbanization started to threaten the Musashino Uplands. In 1914, the Tojo Railway opened between Ikebukuro and Tomezawa. Then, in 1915, the Musashino Railway was opened between Ikebukuro and Hannou. Adjacent areas underwent residential construction and, after the Kanto earthquake disaster in 1923, urban institutions, such as academic, industrial, and military institutions, were built (Yamaga 1971). Consequently, by the 1970s, flatland woodlands in the southern Musashino Uplands had disappeared (Figure 3.4).

There is now little remaining satoyama landscape on the uplands. The satoyama landscape in these areas is limited to the Santome Shinden and Heirinji Temple Woodlands, which are under the designation of several greenery conservation acts.

On the other hand, woodlands remaining on steep slopes in the southern Tama Hills survived comparatively well until the 1960s. This was due to the difficulty of grading the land but, with new grading technology, large-scale projects destroyed most of the remaining satoyama landscapes to create large residential suburbs. The only remaining satoyama landscapes are conserved in Jike Furusato Village and Maioka Park.

However, flatland woodlands in the Musashino Upland that continued to provide fuel and fertilizer after new agricultural field development in the 17th to mid-18th century were mostly destroyed by the 1970s due to urbanization. There were some planned developments by public bodies or big investors, but the main cause of urban sprawl was small developments on former farm plots.

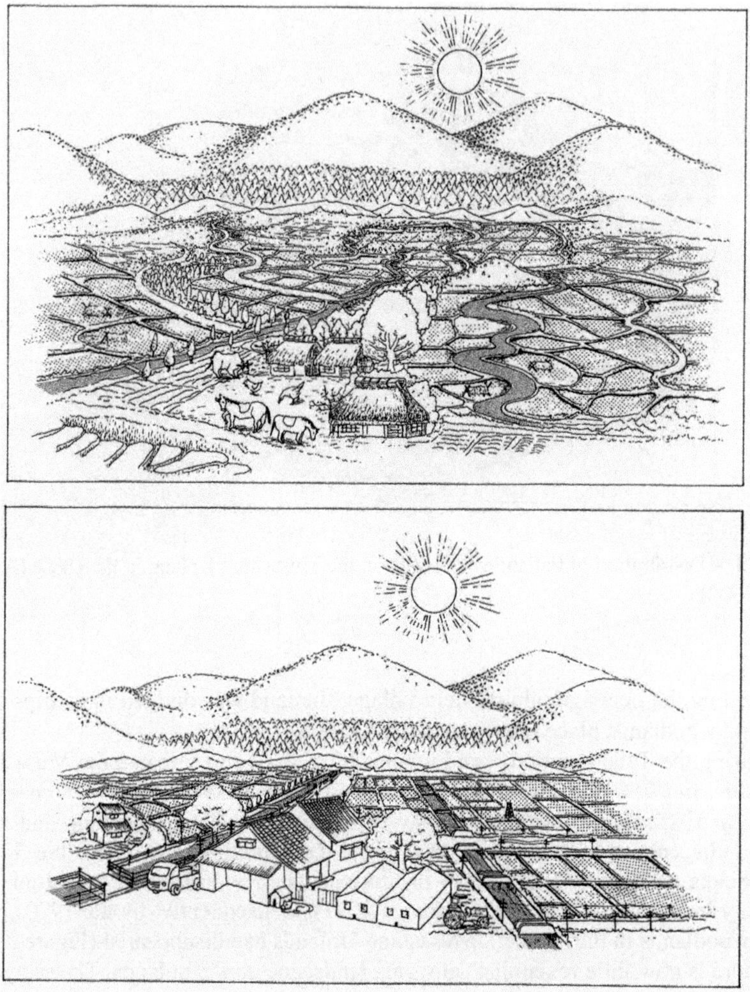

Fig. 3.5. Changes in the relationship between humans and satoyama landscapes. Upper figure: a satoyama landscape from the Edo period to circa 1955. Lower figure: a satoyama landscape since 1965. In former satoyama landscapes, a coppice produced fuel (wood and charcoal) and fertilizer (fallen leaves), and *kayaba* (grassland) produced fodder for livestock. Farmland produced crops and vegetables, and their by-products were used for livestock fodder and fertilizer. In this way, people's lives strongly depended on satoyama landscapes, which provided food, fuel, fertilizer, fodder, and building materials. Recently, electricity, gas, and petroleum have been used as energy sources instead of wood fuel and charcoal, and chemical fertilizers have been used instead of organic compost. Building materials have been diversified and most are imported. Thus, the traditional role of the satoyama landscape is decreasing

3.1.4 Relationship Between Satoyama Landscapes and the People

Hill satoyama landscapes and upland satoyama landscapes have followed different processes over time. The common factor is that they are both losing their relationship with local farmers.

As shown in Figure 3.5, for almost 400 years from the early modern age to the 1960s, farmers depended largely on local bioresources. That is, carbon dioxide from the atmosphere and water from rain were transformed into organic substances by solar energy, and transformed by plants into chemical energy. Products from this so-called *plant factory* utilizing natural energy were food (such as vegetables and crops), fodders, fertilizers (such as fallen leaves or crop litter), and building materials.

After the 1960s, the fuel and fertilizer revolution, the strong yen, free trade, and the import of foreign wood material and fodder prompted many farmers to abandon their coppice woodlands. Consequently, satoyama landscapes have lost their traditional role as suppliers of fuel, fertilizer, fodder, and building materials.

However, satoyama landscapes do not function only in this regard. They also have a function as a habitat for wildlife and a place for recreation, providing people with places for quiet contemplation and reflection, as well as being a repository of traditional culture.

Urban sprawl has brought urban citizens closer to satoyama landscapes. Accordingly, satoyama landscapes and urban citizens need to develop a new relationship. It is important to realize the meaning and value of the satoyama landscape in its modern context.

3.2 Satoyama Landscape Transition in the Kanto Area

A. Tsunekawa and T. Bessho

3.2.1 A Two-Dimensional Analysis of Satoyama Landscape

As discussed in the previous section, despite the profound economic and lifestyle changes since the Meiji period, there has not been a significant change in the area of woodland and agricultural land in Japan. However, the details reveal there was a big change around the cities, where population inflow and urban expansion transformed farmlands and woodlands into residential suburbs. Since the Meiji period, interior woodlands were the focus of conservation, whereas satoyama landscapes distributed in urban surroundings steadily declined. This was especially so in Tokyo and the surrounding Kanto Plain.

The transition of satoyama landscapes over time was analyzed through the use of a Geographic Information System (GIS). The GIS allows analysis of when, where, and in relation to what other incidents changes have happened.

Two-dimensional analysis of satoyama landscape distribution requires geographic information about the satoyama landscapes themselves. For example, the Japanese Ministry of the Environment has developed a natural environment GIS dataset that is open to the public. It includes a 1 : 50 000 digital vegetation map from the national survey on the natural environment, known as the national green census. It is now not too difficult to create one's own vegetation map of a large area by combining aerial photos, satellite images, and field data.

But what about the geographic information of the past? Color aerial photos date from around 1974, black and white aerial photos date from around 1961, and LANDSAT images date from 1972. In 1946, the United States military took black and white aerial photographs. Before this time, aerial photos and satellite images were not available, so geographic information can only be estimated from topography maps, statistical information from political districts, and descriptive literature.

In recent satoyama landscape research, a *jinsoku* (quick survey) map, which was created in the early Meiji period, has often been used. Jinsoku maps were the first topography maps in Japan surveyed using modern methods. The Japanese army of the time made such maps of the Kanto area. Notes attached to the maps help understand the vegetation of the time (Ogura 1993, 1994; Masubuchi and Uenishi 1997).

A jinsoku map is rendered in different colors for each land use: woodland, crop field, rice paddy field, river, road, grassland, and so on. The vegetation types within

Fig. 3.6. Original *Jinsoku* (Quick Survey) Map of the area around Sakahama, Inagi City, Tokyo. The darker parts represent woodland. Japanese characters are vegetation symbols

woodlands, as shown in Figure 3.6, are only described in words, and it is not possible to distinguish between different plant communities. Because of this difficulty, the method used thus far has been to represent vegetation by the most often described vegetation type in any given area of the map sheet (Ogura 1994). However, this method ignores important spatial information of location.

The authors have made an attempt at a detailed analysis of satoyama landscape transition from the early Meiji period until today. Jinsoku maps were used, with due care to lose as little information as possible. The subject area was the Tsurumi River basin in the Tama Hills.

3.2.2 Spatial Analysis of Satoyama Landscape Transition

The jinsoku maps were the main source of information on vegetation in 1880. Information for other time periods was acquired from the following sources: a 1 : 25 000 vegetation map (Kanagawa Prefecture Education Board 1972); a 1 : 25 000 vegetation map of Tokyo Metropolis (Tokyo Metropolis 1970); and a vegetation map (digitized data from a 1 : 25 000 original map; 1990). Topographical data were gathered from the Landform Classification Map included in the Land Classification Map of Tokyo Metropolis and Kanagawa Prefecture, and 50 m grid altitude data from the former Japanese National Land Agency (now a part of the Ministry of Land, Infrastructure and Transport). Slope maps were created from elevation data.

Land use boundaries and vegetation descriptions on the jinsoku maps were scanned into the computer. These raster data were converted to vector data with TNTmips (GIS software; Microimage Inc., Lincoln, Nebraska, USA). On the jinsoku maps, individual houses have been drawn in, but areas where houses are clustered were input as villages in this study in order to give a village land use legend.

The Landform Classification Map was digitized as polygon data. For efficiency, only the inside area of a 100-m radius circle from the center of each vegetation description letter on the jinsoku maps was digitized.

3.2.2.1 Vegetation Classification

As mentioned previously, there was no detailed vegetation boundary on the Jinsoku map. Vegetation types had only been described in letters. Therefore, chronological spatial analysis by overlaying maps was difficult. Consequently, the following method was used to examine vegetation transition over time. A 10-mm radius circle (100 m on the ground) was drawn from the center of each vegetation letter on the map (Table 3.2). Only vegetation types of 1970 and 1990 within these circles were examined for comparison. In the case of more than two vegetation types included in a circle, the vegetation type of the largest area was considered. Table 3.3 shows the correlation for the description of vegetation types. Forest and woodland was classified into five types: Secondary woodland of deciduous oaks (*Quercus acutissima*, *Quercus serrata*), Secondary woodland of *akamatsu* or Japanese red pine (*Pinus*

Table 3.3 Vegetation classification

This study	Rapid survey map	Vegetation map in the 1970s (Tokyo)	Vegetation map in the 1970s (Kanagawa)	Vegetation map in the 1990s
Secondary woodland of *Quercus acutissima*, *Quercus serrata*	*Quercus serrata*	*Quercetum acutissimo-serratae*	*Quercetum acutissimo-serratae*	*Quercetum acutissimo-serratae*
Secondary woodland of *Quercus acutissima*, *Quercus serrata*	*Quercus acutissima*	*Quercetum acutissimo-serratae*	*Quercetum acutissimo-serratae*	*Quercetum acutissimo-serratae*
Secondary woodland of *Quercus acutissima*, *Quercus serrata*	Other deciduous trees	*Quercetum acutissimo-serratae*	*Quercetum acutissimo-serratae*	*Quercetum acutissimo-serratae*
Secondary woodland of *Pinus densiflora*	*Pinus* spp.	*Pinus densiflora* plantation	——	——
Plantation of *Cryptomeria japonica*, *Chamaecyparis obtusa*	*Cryptomeria japonica*	*Cryptomeria japonica*, *Chamaecyparis obtusa* plantation	*Cryptomeria japonica*, *Chamaecyparis obtusa* plantation	*Cryptomeria japonica*, *Chamaecyparis obtusa* plantation, *Chamaecyparis pisifera* plantation
Plantation of *Pinus thunbergii*	——	——	*Pinus thunbergii* plantation	*Pinus thunbergii* plantation
Evergreen broad-leaved forest of *Quercus myrsinaefolia*	——	*Quercetum myrsinaefoliae*	——	——
Secondary grassland of *Miscanthus sinensis*	——	*Arundinario chino—Miscanthetum sinensis*	*Arundinario chino—Miscanthetum sinensis*	*Miscanthion sinensis*
Roadside	——	——	——	Roadside weed community
Dry farm	——	Field weed community	*Pinellio ternatae–Euphorbietum pseudo-chamaesyesis*	Field weed community
Paddy field	——	Paddy field weed community	*Sagittario-Monochorietum*	Paddy field weed community
Paddy field	——	Weed community in uncultivated paddy field	——	Paddy field weed community
Deciduous orchard	——	Deciduous orchard	Deciduous orchard	Deciduous orchard

This study	Rapid survey map	Vegetation map in the 1970s (Tokyo)	Vegetation map in the 1970s (Kanagawa)	Vegetation map in the 1990s
Other land use	———	Artificial lawn	*Zoysion japonicae*	Artificial grassland
Other land use	———	Artificial meadow	*Zoysion japonicae*	Artificial grassland
Other land use	———	Residential district with many planted trees	Residential district with many planted trees	Residential district with many planted trees
Other land use	———	Urban and residential district with few green spaces	Urban and residential district with few green spaces	Urban and residential district with few green spaces
Other land use	———	Land constructed for residence and factory etc.	Land constructed for residence and factory etc.	Land constructed for residence and factory etc.

densiflora), plantations of *Cryptomeria japonica*, *Chamaecyparis obtusa*, Evergreen broad-leaved forest of *Quercus myrsinaefolia*, and Plantation of kuromatsu or Japanese black pine (*Pinus thunbergii*). Other areas were classified as *Miscanthus sinensis* grassland, roadside weed, dryfield, paddy field, deciduous orchard and urban areas. Recreational grassed areas, including golf courses, were classified as urban areas. The eight categories of land use polygon data were: woodland, village, grassland, paddy field, dry field, open water, other, and unclassified.

3.2.2.2 Landform Classification

The area analyzed is mostly hills and valley bottoms, except for the Sagamihara Upland in the south-west. Geomorphic classifications given for the points (based on locations of the vegetation description letters) in this area were classified as uplands. A contour map was created from elevation data. Elevation data, slope data, contour lines, and point data (location of vegetation description letters) were all overlaid. Each point was classified either as a hill slope or a terrace surface.

According to Tamura (1981), there are micro-landform units in valleys: crest slopes, side slopes, head hollow, head floors, bottomlands, and channel ways. Tamura (1981) also pointed out that side slopes were subdivided by the distinct convex break of slope or convex microsegment into two categories of upper side slopes and lower side slopes. This classification followed a procedure using elevation data, slope data, and contour lines. The break line was drawn first. Slopes above the line are upper side slopes, whereas slopes below the line are lower side slopes.

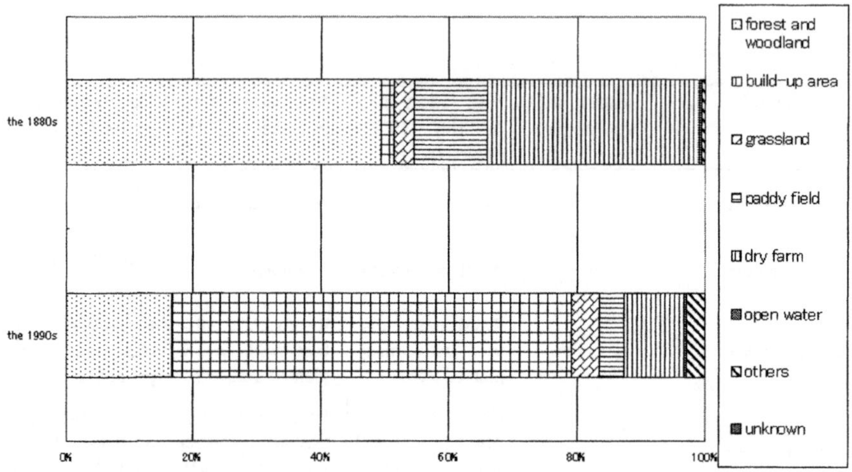

Fig. 3.7. Land use composition in the 1880s and 1990s

3.2.3 Satoyama Landscape Transition from the Early Meiji Period

3.2.3.1 Land Use Transition

The land use composition of 1880 and 1990 is shown in Figure 3.7. Woodlands covered 49% of the area in 1880, and this had decreased to 17% by 1990. Dry fields and paddy fields had decreased by one-third. In contrast, urban areas had expanded greatly, spreading from 2% of the entire area in 1880 to 62% in 1990.

3.2.3.2 Woodland Transition

The species composition of woodlands of 1880, 1970, and 1990 is shown in Figure 3.8. In this figure, the values for 1880 are given as percentages of the number of plant descriptions that appeared on the jinsoku maps, such as *Quercus* and *Pinus*. The values for 1970 and 1990 are the area composition of the woodlands. Seventy-one percent of woodlands was *Quercus* woodlands in 1880, which increased to 83% in 1970 and 91% in 1990. *Pinus densiflora* woodland, which covered 27% of the area in 1880, accounted for only 1% in 1970 and had disappeared completely by 1990. *Chamaecyparis obtusa* and *C. japonica* plantations accounted for 2% in 1880, 10% in 1970, and 7% in 1990. These plantations are now second in area only to *Quercus* woodland.

Next, we followed land use transition of some points that were woodland in 1880. Figure 3.9 shows the result in diagrammatic form. From 1880 to 1970, 46% of deciduous *Quercus* woodland (consisting primarily of *Q. acutissima* and *Q. serrata*) remained and 2% changed to evergreen *Quercus* (consisting primarily of

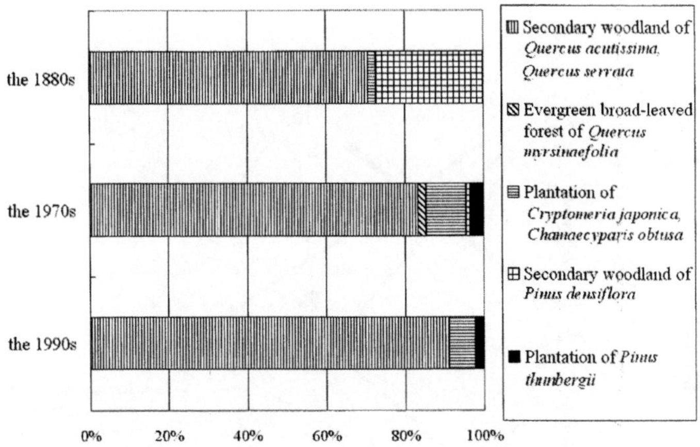

Fig. 3.8. Vegetation composition of forests and woodlands

Quercus myrsinaefolia) forest. On the other hand, only 2% of *P. densiflora* woodland remained, whereas 40% changed to deciduous *Quercus* woodland. Other major transitions were as follows: 6% of deciduous *Quercus* woodlands changed to *C. obtusa* and *C. japonica* or black pine plantations; and 8% of *P. densiflora* woodland changed to *C. obtusa* and *C. japonica* plantations. Overall, the plantation areas have expanded.

Woodlands also changed to other land uses, as shown in Figure 3.10. Seventy-three percent of deciduous *Quercus* woodland and 67% of *C. obtusa* and *C. japonica* woodland have been developed into urban land use. In contrast, only approximately 54% of *P. densiflora* woodland has been developed into urban land use, whereas much of it has been transformed into agricultural land use, with 27% being converted into dry fields and 12% being converted into paddy fields.

The percentage of deciduous *Quercus* woodland remained at 36% through 1970 and 1990. Most of the *C. obtusa* and *C. japonica* and *Pinus thunbergii* plantations that existed in 1970 were transformed into urban areas or into deciduous *Quercus* woodland.

Thus, in the area analyzed, whereas the total woodland area had decreased, the percentage of land of deciduous *Quercus* woodland has increased. *Pinus densiflora* woodlands have decreased, due, in part, to them being taken over by *C. obtusa* and *C. japonica* plantations. The number of points that were located in woodland decreased by 50% from 1970 to 1990. This indicates that destruction of woodlands in satoyama landscapes continued after 1970.

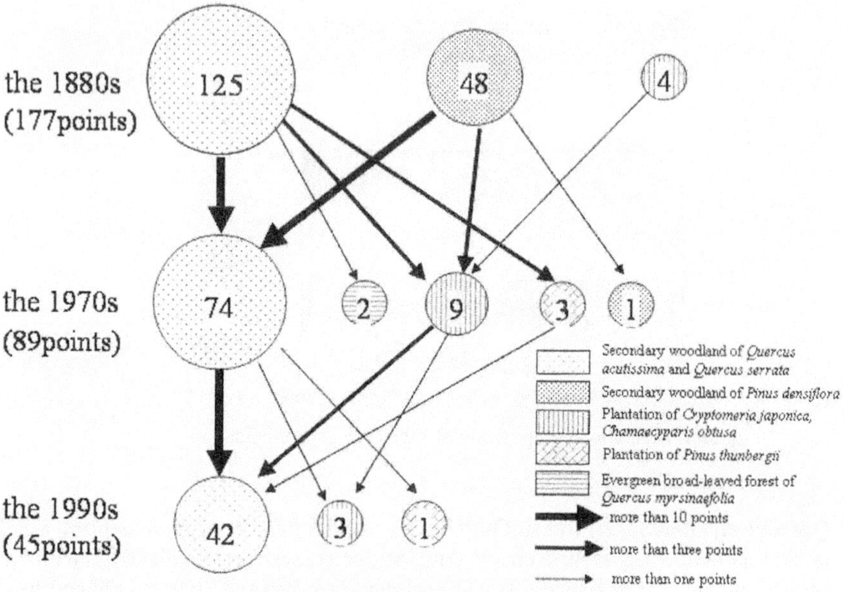

Fig. 3.9. Changes in wooded land

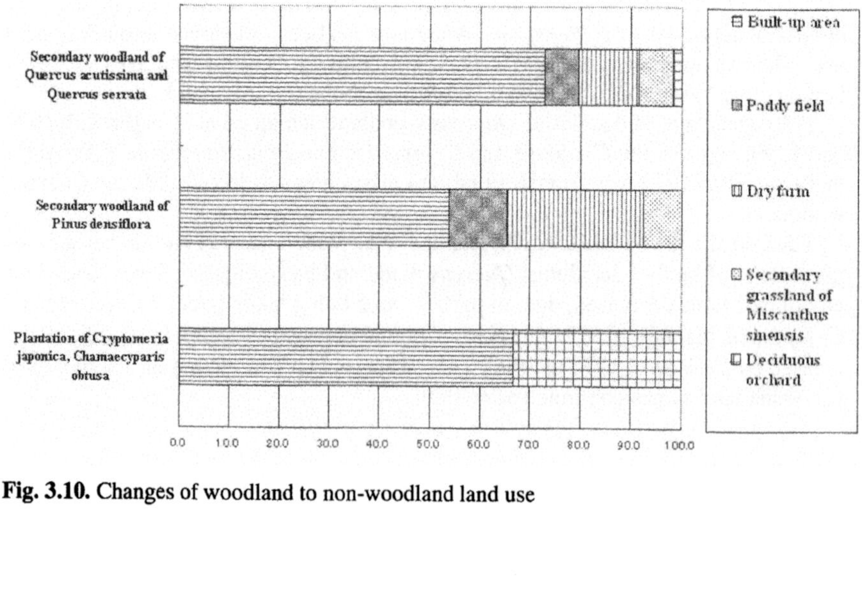

Fig. 3.10. Changes of woodland to non-woodland land use

3.2.4 Causes of Satoyama Landscape Transition

The satoyama landscapes analyzed have undergone a great transition in terms of vegetation and land use from the early Meiji period to today. Woodland covered 49% and agricultural land (dry and paddy fields) covered 15% of the area in the early Meiji period, but these figures had decreased to 17% woodland and 8% agricultural land by 1990. Woodlands were mainly taken over by urban areas. The remaining woodlands were concentrated on hills and terrace scarps. Particularly until 1960, woodland in the hills remained due to difficulties of construction in these areas, although as technology improved in the 1960s, hills became a target for residential development. It is no longer realistic to expect satoyama landscape woodlands to remain based only on geomorphic reasons. Satoyama landscapes will probably continue to disappear unless conservation action is undertaken. These actions could include restrictions of land use for the designation of greenery conservation areas, urbanization control areas, or public municipalities conserving satoyama landscapes as parks.

Qualitatively, deciduous *Quercus* woodlands have always been dominant throughout the period in question. *Quercus* woodlands have increased from 71% to 91%, whereas *P. densiflora* woodlands have disappeared in 1990 from comprising 27% in 1880. Sixty-nine percent of *P. densiflora* woodland has been turned into urban areas, and 17% has been converted to deciduous *Quercus* woodland. Transformation from *P. densiflora* to *Quercus* woodland is also apparent. In relation to landform, transformation from *P. densiflora* to *Quercus* woodland often occurred on lower middle slopes (Figure 3.11). In nature, *Pinus* grows well on hill tops because it prefers dry soils. It is assumed that because farmers quit actively using *Pinus* woodlands, these woodlands have been replaced by competitive species, such as deciduous *Quercus*. One aspect of the rapid decrease of *P. densiflora* woodland

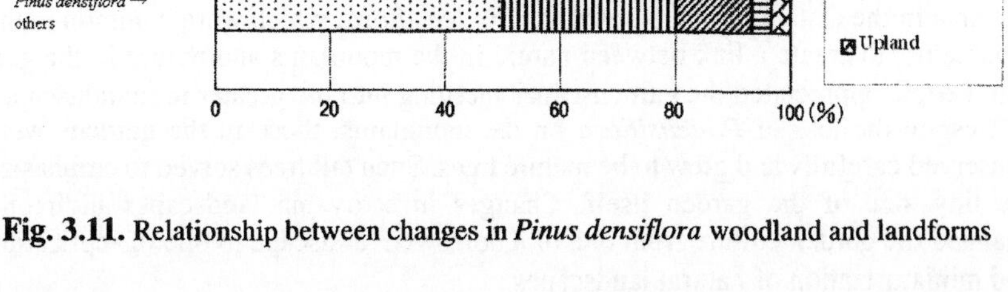

Fig. 3.11. Relationship between changes in *Pinus densiflora* woodland and landforms

that cannot be ignored would be pine wilt disease, which has expanded since the 1960s.

The satoyama landscape analyzed went through a great transition in both quantity and quality. It is expected that various and effective conservation criteria will be needed in order to conserve satoyama landscapes, as well as adequate management with good consideration of the natural condition of the area.

3.3 Satoyama Landscape Transition in the Kansai Area

J. Morimoto and Y. Morimoto

3.3.1 Sanzan

While the satoyama landscapes of the Kanto area are characterized by *Q. serrata* and *Q. acutissima* woodlands, those in the Kansai region were long characterized by *P. densiflora* woodlands. In fact, some parts of Higashiyama (eastern mountains), Kitayama (northern mountains), and Nishiyama (western mountains), which are collectively called *Sanzan* (the three mountains), were covered with beautiful *P. densiflora* woodlands until quite recently. These *P. densiflora* woodlands have been destroyed by pine wilt disease and have been replaced by laurel forests.

Looking at the hills around Kyoto today, it may be hard to believe that the satoyama landscapes in Kansai were covered with azaleas during the Meiji period (1867–1911) instead of the present *C. obtusa*. Investigations over the past few decades have revealed a dynamic transition in species composition.

The *P. densiflora* woodland of Arashiyama, a part of Nishiyama, was at its most beautiful when Humiaki Okazaki researched and published *Woodland Amenity and Recreation* in 1972. The beauty of the old *P. densiflora* trunks remained in people's minds of the time. Before 1972, the *P. densiflora* there were short and shrub-like, but they then disappeared completely and were replaced by laurel forests, consisting primarily of *Quercus glauca*.

What about the *P. densiflora* in Higashiyama? They were not mature woodlands. In the Modern and Meiji periods, gardens designed by Ueji were the essence of culture in Kyoto (e.g. gardens such as Nomura's villa garden and Hosokawa's villa garden around Higashiyama). Gardens in Kyoto share the characteristics of having flowing water (from Lake Biwa) and of borrowing the landscape of Mount Higashiyama in the distance. Ueji planted thin *P. densiflora*, which were common in the mountains, to create a link between nature in the mountains and nature in the garden. People appreciated the gardenscapes merging into the greater mountainscape.

Despite the loss of *P. densiflora* on the mountains, those in the gardens were conserved carefully and grew to be mature trees. Such tall trees served to emphasize the tiny size of the garden itself. Changes in satoyama landscapes indirectly changed the garden culture from one of a borrowed landscape to one of replication and miniaturization of natural landscapes.

The hills of the Paleozoic and Mesozodic formations in the Kitayama area, except for the woodland areas of *C. japonica*, also suffered from pine wilt. There are various types of woodlands in the area, but the dominant species is *C. obtusa*. *Chamaecyparis obtusa* woodlands become intensely dense as they grow, resulting in a very dark understory that will not allow invading species to grow. The lack of an understory community can, however, cause soil erosion, and it is possible that the lack of understory stops natural succession.

Thus, *P. densiflora* woodlands that used to characterize the Kyoto landscape have been transformed into various other types of woodland over a relatively short period of time.

3.3.2 Satoyama Landscapes with Pine and Azalea Woodlands

The landscape of the mountains surrounding the Kyoto Basin started to transform from virgin forest into secondary *P. densiflora* woodlands in the late Heian period (10th century). By the middle of Meiji period, the mountains were covered with *Pinus* woodlands, shrubs, and grasslands. Denuded mountaintops were not uncommon. Ogura (1992) translated old maps of Kyoto by carefully excluding subjective aspects of the drawings. It is probable that, through over-extraction during the Edo period (17th–19th century), woodlands around Kyoto were shrubby rather than sparse *Pinus* woodlands.

A more objective discussion is possible on the woodland condition of the mid-Meiji period. According to Ogura (1996), the surrounding mountains were covered mostly with *P. densiflora* ranging in height from 2 to 5 m, whereas Mount Hieizan was covered with shrubs approximately 1.5 m in height (Figure 3.12). A resident of what is now Kyoto University sang of '*kurenai moyuru oka no hana*' (the blooming red flowers on the hills). This is about the wild azaleas on Yoshidayama, isolated hills adjacent to Mount Higashiyama. Today, this area is a dark laurel forest and is more likely to be described as dark green woodlands on the hills. The azaleas are losing their significance. There were many *Rhododendron macrosepalum* that favored the fine-textured soils of the Osaka formations. On the other hand, deciduous *Rhododendron reticulatum* was dominant on hard chert bedrock areas, which are said to have developed during the Triassic Period of the Mesozodic formation, or on the thin soil of granite gravels. Other than these, *Rhododendron kaempferi*, with red flowers, and *Rhododendron tectum*, which is a natural hybrid between *R. kaempferi* and *R. macrosepalum*, were the major wild azaleas in the area.

These azaleas were well suited to coppicing and used to be one of the major producers of firewood. Azaleas in Narabigaoka, which is another series of isolated hills in the west of the Kyoto basin, were also well known. However, there is no evidence of tall trees in this area. Wild azaleas in this area were mostly *R. macrosepalum*.

It is difficult to determine precisely when these woodlands disappeared. From paintings, maps, and other materials, it seems likely that satoyama landscapes in

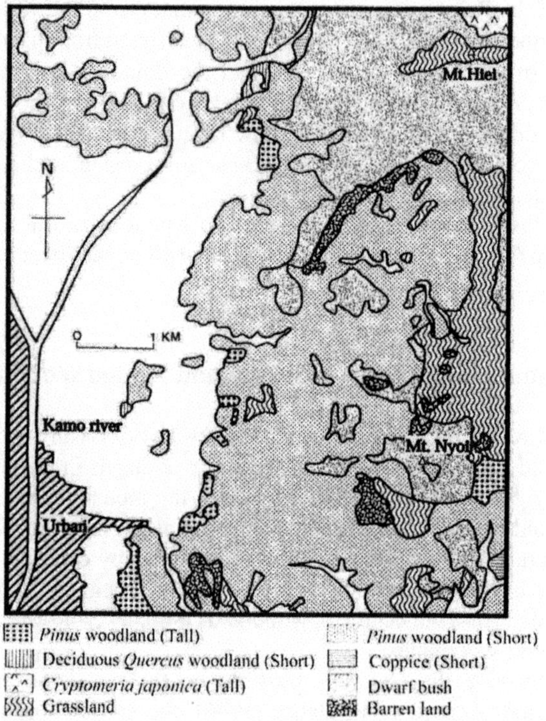

Fig. 3.12. Vegetation map for the north-eastern mountain district of Kyoto in the middle of the Meiji period (1868–1911). (After Ogura 1996). Blank areas are mainly agricultural land

Kyoto were degraded woodlands due to firewood extraction rather than through the intentional development of mature *Pinus* woodlands.

During the Showa period (20th century), satoyama landscapes surrounding Kyoto were transformed dramatically. Table 3.4 shows the affect of the Muroto Typhoon of 1934, which drastically damaged not only the woodlands around Mount Higashiyama, but also the Tadasu No Mori Forest. A forest of more than 1000 tall trees was reduced to less than 100 trees. The woodlands on Mount Higashiyama behind Kiyomizu Dera Temple were almost totally destroyed. This was the only place in Kyoto that had mature woods. Old photographs show the dreadful sight of uprooted *Chamaecyparis obtusa* and *Castanopsis sieboldii* uprooted. *Pinus densiflora* was not uprooted because the roots of these trees go down into the bedrock. Instead, they split in the middle.

Castanopsis sieboldiii and *Myrica rubra* were planted later, which is a possible cause of the transformation into laurel forest. Even in Tadasu No Mori Forest, *Cinnamomum camphora* were planted in the Shimogamo Shrine as a gift from the

Table 3.4. Phenomena that have seriously affected woodland physiognomy in Kyoto City

1934	Muroto typhoon causes severe damage to Higashiyama Restorative tree planting follows damage
1940s–1950s	Indiscriminate cutting trees in the chaotic era after World War II
1960s	Increased use of fossil fuel leads to a sharp decline in the use of firewood and charcoal
1970s	Pine wilt disease causes extensive damage
1980s	Succession of *Pinus* woodlands damaged from pine wilt disease

Japanese Department of the Interior of the time, although they are not natural vegetation. It is not clear why such laurel trees were favored at the time.

The confusion that followed World War II had a major impact on satoyama landscapes. Statistical documents show that firewood and charcoal production decreased from 90 000 m^3 in 1910 to 40 000 m^3 in 1930, whereas 60 000 m^3 was produced after the war. Coniferous building material production jumped from 10 000 m^3 in 1930 to 40 000 m^3 in 1950. On the other hand, charcoal production, which was at its peak in 1920, continued to decrease as a result of the fuel revolution. In 1965, charcoal production was less than 20% of the production in 1955.

As well as losing their function as coppice woodlands, satoyama landscapes in Kyoto went through several phases of natural and human disturbance.

3.3.3 Woodlands in North-eastern Kyoto After Pine Wilt Disease

The reality of the destruction of the *P. densiflora* woodland and the subsequent succession cannot be described be adequately by the simple saying that it became laurel forest. Figures 3.13 and 3.14 are the results of investigations over a period of 20 years. The investigation was performed for the period from 1974 to 1995 in the north-eastern woodlands, including the national forests on Mount Higashiyama (Morimoto and Yoshida 2000).

That study revealed that the *P. densiflora* woodlands of the satoyama landscapes in the Kansai area were not dominant in 1974. The Experimental Forest of Kyoto University (in Kamigamo in Kitayama) was *P. densiflora* woodland on barren soil when the United States military cleared the area after World War II. Taking advantage of the conditions, the university collected *Pinus* and azalea specimens but, by 1974, *C. obtusa* was excessive. As mentioned earlier, this is a typical pattern of satoyama landscapes in the Kansai, where *C.obtusa* regenerates naturally.

Chamaecyparis obtusa is tolerant of barren or thin soil. Its shade tolerance is also very high and it has a long lifespan. These characteristics allow *C. obtusa* to develop enclosed woodlands. When constructing the Fujiwara Kyo capital, extraction of *C. obtusa* from nearby woodlands is said to have been a cause of the denudation of Mount Tanakamiyama. It is commonly thought that the natural climax forest of this area is laurel forest. However, on granite or chert soils, or on cliffs, *C. obtusa*

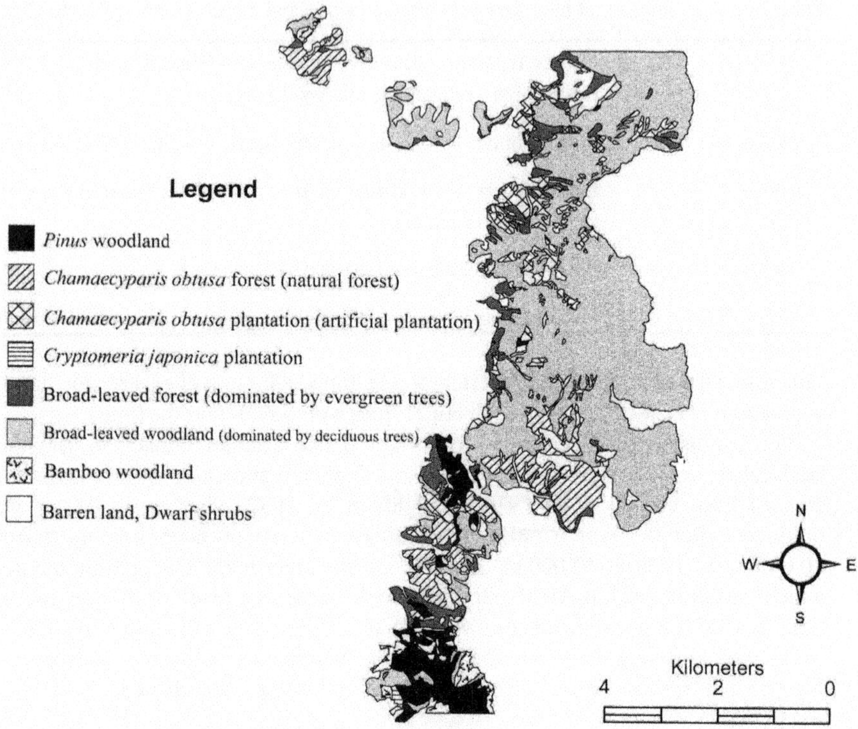

Fig. 3.13. Vegetation map for the north-eastern mountain district of Kyoto in 1974 (Morimoto and Yoshida 2000)

forest is likely to be the edaphic climax. Granite weathers quickly and this, along with erosion, limits soil formation. Chert, however, weathers more slowly. Hozukyo, which runs through Arashiyama to Kameoka, has steeply sloped ground of this chert bedrock, where azaleas grow wild.

Deciduous broad-leaved woodlands covered the greatest area in 1974, although these woods were not mature. In the mid-Meiji period, they were shrubs. From Mount Daimonjiyama to Mount Hieizan, the soil is unstable. Heavy rains often cause floods, erosion, landslides, and debris avalanches. It is probable that human interference, such as firewood gathering, hampered forest development.

Dark rock, called hornfels, is granite hardened by magma heat. This is what keeps the peaks of Hieizan and Daimonji from eroding. The hard structure of hornfels does not weather easily and, thus, soil development is slow. It is difficult for a forest to develop once the surface soil is gone. Evidence suggests that, during the mid-Meiji period, there were quite a few naturally occurring grasslands and denuded mountaintops.

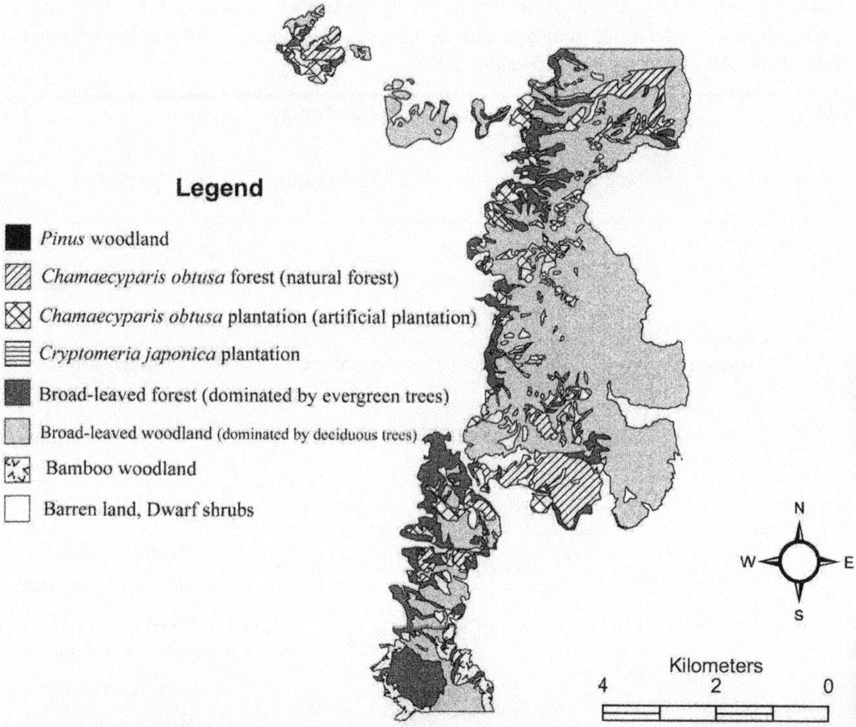

Fig. 3.14. Vegetation map for the north-eastern district of Kyoto in 1995 (Morimoto and Yoshida 2000)

Pinus woodlands had disappeared by 1995 and laurel forest had increased in area almost 2.5-fold.

3.3.4 Wild Azaleas After Pine Wilt Disease

When the *P. densiflora* woodlands disappeared, what happened to the wild azaleas? Wild azaleas grow in *P. densiflora* woodlands and *Q. serrata* woodlands but, as laurel forest takes over, the landscape of *kurenai moyuru* (red blooms) is certain to disappear. A closer look at the disappearance of azaleas from other environments was described by Morimoto and Yoshida (1999a, Morimoto 2000). Have azaleas really disappeared, like *P. densiflora*, or are they just hidden under thick woodlands?

Wild azaleas are long-lived trees. Their growth phase can be divided into three periods derived from form and propagation ability. Those phases are as follows: the prereproductive phase, which is before blooming; the reproductive phase; and the post-reproductive phase, when the azalea loses its breeding potential. The *Q. ser-*

Table 3.5 Landscape types and survival of native species of *Rhododendron* based on the combination of vegetation, geology, and topography of modern satoyama landscapes of Kyoto City (After Morimoto and Yoshida 2000)

No.	Landscape types (ecotopes)			
	Surface geology		Topography	Vegetation
1	Unconsolidated sediment	(conglomerate, sandstone, mud-stone)	Flat land	Bamboo woodland
2		(Osaka group)	Slope	Evergreen forest
3		(Osaka group)	Slope	Deciduous woodland
4	Mesozoic and Paleozoic	(C, S)	Ridge	*Pinus* woodland
5		(C, S)	Ridge(upper)	Natural forest of *Chamaecyparis obtusa*
6		(C, S)	Ridge(lower)	Natural forest of *Chamaecyparis obtusa*
7		(C, S)	Slope	Natural forest of *Chamaecyparis obtusa*
8		(C, S)	Valley(upper)	Natural forest of *Chamaecyparis obtusa*
9		(C, S)	Valley(lower)	Natural forest of *Chamaecyparis obtusa*
10		(C)	Valley	Artificial plantation of *Cryptomeria japonica*
11		(C, S)	Slope	Evergreen forest
12		(C)	Ridge	Deciduous woodland
13		(C)	Slope	Deciduous woodland
14		(C, S)	Valley	Deciduous woodland
15	Granite		Slope	Artificial plantation of *Chamaecyparis obtusa*
16			Valley	Artificial plantation of *Cryptomeria japonica*
17			Slope	Deciduous woodland

Surviving native species of rhododendron					
R. reticlatum	*R. macrosepalum*	*R. obtusum*	*R. tectum*	Population (per ha)	Juvenile population (per ha)
o	o	-	-	311	0
-	o	-	-	533	267
-	o	-	-	133	133
o	o	-	-	1200	733
o	o	-	-	800	250
o	o	-	-	1267	700
o	o	-	-	444	0
o	o	-	-	2933	752
-	-	-	-	-	-
-	-	-	-	-	-
o	-	o	-	2978	2978
o	-	-	-	2533	507
-	o	-	-	311	272
-	-	-	-	-	-
-	-	-	-	-	-
-	o	o	o	7067	7009

rata forest, the *C. obutusa* forest in Takaragaike, and the *Q. glauca* forest in Kamigamo, which are different successional stages of satoyama landscapes, have all been studied. Canopy growth rates and proportion were analyzed, and future change was predicted from correlating light conditions and branch growth data, as well as light conditions and blooming data, for wild azaleas in forests in different stages of succession.

That study revealed that azaleas grow slowly but steadily at the forest edge, while their canopy growth corresponds to subtle light conditions in the interior of the forests. Their canopies grow larger in lighter environments, but smaller with shorter branches in darker environments. The study also revealed that wild azaleas in the *Q. glauca* forest were older, whereas those in the *Q. serrata* woodland were younger and more slowly growing.

Aerial photographs and remote sensing are useful for forest canopy inventory, but not for understory inventory. Forest canopy inventory has been used to describe forest types in this analysis. Because different land conditions result in different azalea growth, even within the same forest type, influential factors, such as surface soil and land formation, are overlaid on top of forest types, and the areas classified as ecotopes. On the premise that azalea growth tendency is the same within each ecotope, change for the entire area is estimated (Morimoto and Yoshida 2000).

Table 3.5 shows the correlation between ecotope type and wild azaleas. Analysis reveals the trend of azaleas in the Higashiyama area of Kyoto as follows. The overall population has decreased over the past 20 years and surviving individuals have increased slightly in size. Aging is evident from the propagation potential of the azaleas. The red bloom landscape has certainly decreased due to pine wilt and plant succession to laurel forests, but the process is unlike that of the *P. densiflora* woodlands. The red bloom landscape did not decrease sharply and there is some still surviving in the dark woodland understory.

Azalea shrubs are still very capable of revival given favorable environmental conditions. In fact, the *C. obtusa* woodland of Mount Ichijoyama has been carved off from the surroundings by illegal development, resulting in a ring of blooming azaleas around the edge. In spring, it looks as if the mountain is wearing a pink headband. Azaleas have regenerated along forest edges that have been created suddenly as a result of development.

3.3.5 Administrative Efforts to Conserve Satoyama Landscapes

Residents of the Kyoto area are culturally and emotionally attached to satoyama landscapes, especially around the city. While the agroforestry function of the satoyama landscape to support paddy field rice production is now defunct, it is valued for its function as greenery surrounding urban areas.

The city, enclosed by an earthen wall called *o-doi*, which was made by *Kampaku* Hideyoshi, the chief adviser to the Emperor, was once called *Rakuchu*, while the outside was called *Rakugai*. Rakugai, especially along the mountain edges, used to be a rural landscape until the Meiji period. In time, the entire basin was urbanized and the developer's hand fell onto the mountains on the outskirts.

In order to conserve the beautiful scenery of Kyoto as a city of green mountains and clean water, and to protect the surrounding green areas from urban sprawl, a scenic beauty policy was enacted in 1930. In that year, the City of Kyoto designated 3400 ha in Higashiyama, Kitayama, and Nishiyama as an area of scenic beauty. This was the first such designation in Japan. The intention was to promote superior development by restricting building area and building form, but such restrictions caused intense arguments at the time. The scenic beauty designation was an epochal event in the history of Japanese greenery conservation policy.

Although the scenic beauty area was eventually increased, the carrot rather than stick approach was insufficient to withstand the pressures of development that occurred throughout the country in the 1960s. In 1966, the Special Legislative Measure related to the Preservation of the Historic Landscape of the Old Capitals was promulgated, which included a land-purchasing system. This was an attempt to freeze the scenic beauty designation area as it was then. This resulted in stronger development restrictions in Arashiyama, Sagano, and Ohara.

In 1970, the City of Kyoto enacted the City Ordinances Governing Landscape Areas, and conservation of the historical townscape in the urban area took effect, following the Standard of Scenic Zone Bylaw by the Japanese Ministry of Construction. However, the scenic beauty of Kyoto suffered a spate of illegal developments during the bubble economy (late 1980s). So, the City of Kyoto enacted City Ordinances Governing Urban Landscapes and worked on the Natural Scenery Conservation Plan, which led to City Ordinance Governing the Conservation of Natural Scenery in 1995, and a new bylaw for scenic beauty conservation (Abe 1999). The basis for this new bylaw was the Kyoto City Land Utilization and Cityscape Conservation Council, and the report suggests that scenery is characteristic of a place and it well reflects its local culture. Kyoto has created a rich and unique scenery, which is an irreplaceable heritage not only for our nation, but also for the rest of the world (Kyoto City Land Utilization and Cityscape Conservation Council 1995)

Apart from freezing the conservation system as a special zone for the conservation of historic natural features by the Special Legislative Measure related to the Preservation of the Historic Landscape of the Old Capitals, the bylaw aimed to promote development that was compatible with natural and historical contexts. In 1996, the Natural Scenery Conservation Plan was prepared and the designation area was expanded (see Figure 3.15).

The basic idea of the bylaw was to conserve mountain ridges that can be viewed from roads, urban areas, and major villages, and also to conserve, as natural scenery conservation areas, mountain ridges that stretch into urban areas. It excludes historical areas, special greenery areas, land that has already been developed, villages, and cultivated fields. Activities that were subject to restriction were residential development, extraction of earth and woods, piling up objects, and new building construction.

The bylaw not only restricted the height of building structures, as in past bylaws, but also obligated a certain portion of greenery area. At least 70% of natural greenery was to be set aside in the Category 1 Area and 50% in the Category 2 Area. It is of note that a natural rehabilitation guideline was produced in order to conserve

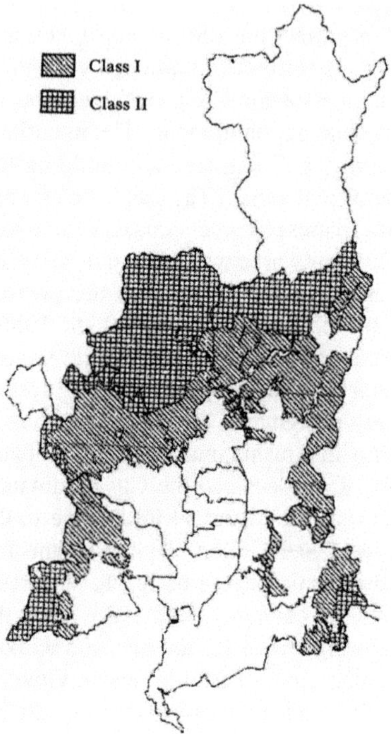

Fig. 3.15. Boundaries of the cities, towns, and villages in Kyoto City and the natural landscape conservation zone 2000

scenic beauty, not just to extend green space. In other words, careful rehabilitation of the vernacularity, diversity, and scarcity of satoyama landscape vegetation were well considered.

There are several opinions about how rehabilitation should proceed. Critics point out the necessity of taking into account the fact that the vegetation in Sanzan itself is changing dynamically and the danger of applying an idealistic new methodology lacking flexibility that may hinder scenic conservation. However, there is no doubt that it is an epochal bylaw in its consideration of quality of greenery. It could be called a *bylaw for conserving satoyama landscapes as borrowed landscape*.

3.3.6 Vegetation Management Technology in Borrowed Landscapes

Although the new bylaw recognized the importance of vernacularity, diversity, and scarcity, protection without management could result in the disappearance of species diversity, as well as the disappearance of Kyoto's culture of borrowing landscape. Research towards developing technology to convert the species composition of woodlands is being conducted in woodlands at the Kamigamo Experimental Forest of Kyoto University, in Kitayama. In order to improve the current situation of s hade-intolerant species disappearing as a result of *C. obtusa* woodlands on the upper slopes and evergreen *Q. glauca* forest on the lower slopes, thinning is being used in an attempt to naturally change the forest into *P. densiflora* and *Q. serrata* woodland. Because total destruction of forests and woodlands is not allowed in the zone for natural amenity conservation, a few old coniferous trees and deciduous trees of at least 5-cm diameter are left, which mimics traditional satoyama landscape management. In areas where trees were removed, conifers sprouted first (Makino et al. 2002). Within 1 year, all local species of trees grew sprouts except *C. obtusa* and *P. densiflora*. However, seed regeneration varied greatly depending on what part of the slope the trees were positioned. On upper slopes, *P. densiflora* and *C. obtusa* emerged well, but only *P. densiflora* survived the summer heat. On middle slopes, *C. obtusa* was in the majority, but some deciduous seedlings, such as *Acanthopanax sciadophylloides* and *Clethra barvinervis*, also grew. On bottom slopes, *Eurya japonica* was in the majority. The expectation is that long-lived seeds of the diverse species that compose *Q. serrata* woodland would germinate successfully, but the results are not yet conclusive.

This project is currently only in its second year, and it is probable that seed regeneration from plants activated by gaps through thinning will occur after the second year. However, if *C. obtusa* and *E. japonica* become dense again, thinning alone may not be effective for species conversion in this case. Flexible management with careful observation is required.

Such development of satoyama landscape management techniques aimed at conserving natural integrity and local landscapes has recently been undertaken across Japan (Kameyama 1996). Satisfying results via long-term, sustainable, and integral means have not yet been achieved. The three mountains in Kyoto are advancing in plant succession to laurel forests and so is the attempt to convert satoyama landscapes from agricultural woodland to woodland of the culture of borrowed landscape. Kyoto is expected to take the lead as a model of sustainable satoyama landscape management.

3.4 Mechanisms of Satoyama Landscape Transformation

M. Yokohari and H. Kurita

3.4.1 Hiki Hills, Saitama Prefecture

In the center of Saitama Prefecture, 30 km north of the Sayama Hills, are the Hiki Hills. If the Sayama Hills are a green island floating in an urbanizing Kanto Plain, the Hiki Hills could be called a green peninsula reaching out from the Chichibu Mountains (Figure 3.16). The Hiki Hills are located on the eastern edge of the

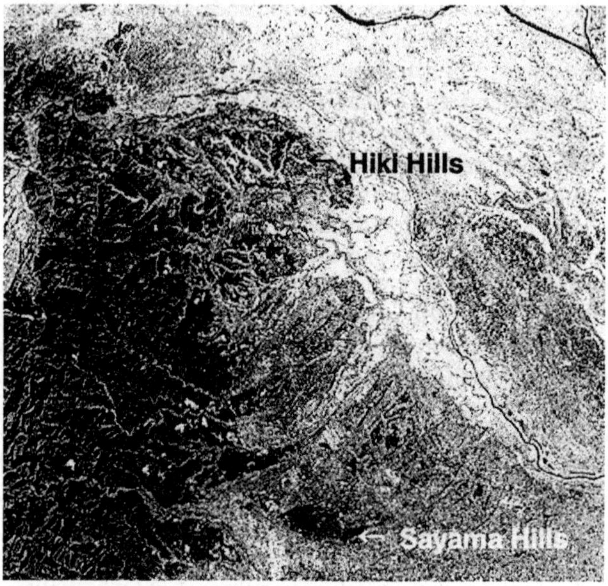

Fig. 3.16. Hiki Hills in Saitama Prefecture

Kanto Plain. Small streams branch through rolling hills to create a landscape of intricate yatsu. At the ends of the valleys there are normally detention ponds for irrigation purposes. The geology of the area is marine clastic rocks formed in the early to mid-Miocene (11 million years – 23 million years BP), called the Matsuyama group, topped with a thin layer of Kanto loam. The soil is mainly yellow-brown forest soil, originating from sandstone, mudstone, and tuff, which are Tertiary, or Kuroboku (Ando) soils originating from volcanic ash.

According to Miyawaki et al. (1976), *P. densiflora* occupies the greatest area in the Hiki Hills because of the dry soil based on the gravel layer. However, in the area, the pine wilt catastrophe since the 1980s left only a few *P. densiflora* woodlands standing. The dominant vegetation at present is *Q. acutissima* and *Q. serrata* woodland. Many of these woodlands are secondary and have the potential to undergo succession to broad-leaved evergreen forest (e.g. *Q. myrsinaefolia* forest) without disturbance. Natural vegetation of the area that can be identified at present includes *Abies firma* forests. On the other hand, in the Hiki Hills, there are endangered species, such as *Tanakia tanago* in Namegawa and *Sasakia charonda* in Ranzan. Both the municipality and citizen groups are actively involved in nature protection.

In this chapter, the transformation of a satoyama landscape to what it is today will be described using the Hiki Hills as an example.

3.4.2 Prewar Satoyama Landscape Transformation

With the development of agricultural technology came the expectation of more efficiency in cultivation, and the nature surrounding croplands started to play a role as a natural supplier of resources, such as fertilizer. There are a few references that describe the conditions of satoyama landscapes in the mid–early modern ages, including Mura Ezu (village illustration maps) of the early modern era, topography maps of the Meiji and Taisho period, and paleographs, such as *kenchi chou* (survey documents) and *mura meisai chou* (village detail documents).

In the Hiki Hills, analysis of satoyama landscapes using illustration maps and old documentation is conducted by way of creating a historic record of the flora by municipalities. The Tamagawa Village Board of Educational (1995) and Tamura (1994) analyzed the transformation of satoyama landscapes in the Hiki Hills using mura ezu and kenchi chou of the early modern ages, and a jinsoku of 1881. From the documents, satoyama landscapes in the Hiki Hills were found to have:

1. Originated from common grasslands called *magusaba*; or
2. Originated from crop fields maintained by fire, which later shifted to coppice woods called *kasegi yama*, with the higher demand of charcoal following the Edo period.

Magusaba supplied green manure, which was indispensable for maintaining land productivity in both paddy fields and dry fields, and also fodder and fire wood (wildly grown shrubs). The magusaba had long been managed by the community as common land. The actual vegetation was mainly herbaceous plants, such as *Miscanthus sinensis*, *M. sinensis* with scattered *P. densiflora*, or *M. sinensis* with scattered young *Q. serrata*. The grasslands of the satoyama landscapes in the mid–early modern age largely consisted of magusaba and kayaba.

Many of the grasslands that were used as common land were sold off in the late Meiji and Taisho periods, divided, and eventually became woodlands due to lack of human disturbance or plantation. Eventually, many of these lands were sold off again, especially after World War II, and were developed into new towns or golf courses. Satoyama landscapes were considered dispensable because of their low productivity and because they were only relatively recently acquired, rather than family property that had been handed down for generations.

On the other hand, coppice woodlands are believed to have originated from *yakihata* (shifting cultivation) that produced foxtail millet and barnyard millet fertilized by burning plants from the mountains in a cycle of a few decades. Such yakihata, which produced food for a long time, was left behind by agricultural technology and the abandoned land became woodlands of *Carpinus laxiflora*. This woodland is completely different from the satoyama landscape woodlands today, which typically consist of *Q. acutissima* and *Q. serrata*. Later, a growing demand for charcoal resulted in selective cutting to create economically beneficial coppice woodlands. The consequence was coppice woodlands of *Q. acutissima* and *Q. serrata*.

Thus, the satoyama landscape in the Hiki Hills underwent dynamic changes through interactions with humans during the medieval and early modern ages.

3.4.3 Post-war Satoyama Landscape Transformation

Quantitative analysis of satoyama landscape transformation after World War II is possible because of the availability of historic records. Aerial photos taken by the United States Army in 1947 are one of the oldest resources that provide detailed and accurate information on land use in satoyama landscapes. Although the photos are in black and white, the images at a 1 : 20 000 scale reveal detailed information. Woodland volume can be calculated using photogrammetry, a method used to gain three-dimensional information from overlapping images of the same area (Figure 3.17).

Aerial photos have been taken approximately every 5 years since 1960. Photos from 1947, 1970, and 1995 were examined. Land use, such as paddy fields, dry fields, construction sites, and woodlands, were identified. For further detail, satoyama landscape vegetation types were identified in four categories focusing on density and volume. Those vegetation types identified were grass, shrub, sparse woodland, and dense woodland (Table 3.6).

Shrub is one of the descriptions written on original jinsoku maps and means woodland that was dominated by *Q. serrata* and *Lespedeza bicolor*. The mountain where the old man went to gather firewood in the popular Japanese children's story *Momotaro* was probably this type of land. It was identified as land occupied by sparse vegetation.

On top of the three photos, a 100 × 100 m grid was overlaid. For each cell, the greatest land cover occupation identified from three-dimensional analysis was plotted to create land-cover maps. A comparison was made for the three time periods (Kurita and Yokohari 2001).

Figure 3.18a is a land-cover map for 1947. Paddy fields and dry fields stand out as the dominant land use. Paddy fields were distributed in yatsu through the hills. In upper areas, dense woodlands, sparse woodlands, shrubs, and grasslands were distributed in a mosaic. Interviews with local farmers revealed that the majority of the area that was identified as grasslands in 1947 was, only a few years prior, *P. densiflora* or *Q. serrata* woodland that was cut down. After such felling, *M. sinensis* grew thick and functioned rather like *kayaba*, place for collecting grasses. Shrub lands were identified as lands at least 5 years after felling, sparse woodlands after approximately a decade, and dense woodlands before the felling of either *P. densiflora* or *Q. serrata* woodlands. Consequently, it can be assumed that the mosaic of these four different land covers indicates different stages in the process of satoyama landscape vegetation management. According to Ide (1992), the characteristic of vegetation in rural landscapes is a mosaic of different stages, both temporally and spatially. It is probable that the four different vegetation types that existed in 1947 were typical components of the spatial mosaic.

Figure 3.18b shows the land cover of the area in 1970. Paddy fields were distributed in the same manner as in 1947, and no significant difference is obvious. In the hills, there are large areas of grasslands: two in the east and one in the north-west. These are golf courses developed after the 1960s. Large-scale developments, such as golf courses and new towns, were the two main causes of satoyama landscape transformation after World War II. For other parts of the hills,

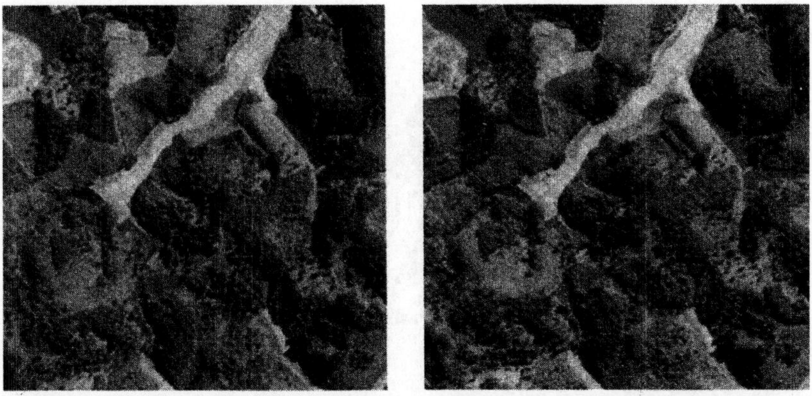

Fig. 3.17. Understanding vegetation in the satoyama by three-dimensional scenes. A three-dimensional scene of a satoyama, identified by pairs of aerial photographs taken at the same site from different angles, such as this pair taken by the United States Army in 1947, may help to understand the vegetation in the satoyama

Table 3.6 Categories of woodland classification

Categories	Identification standard
Grass	Crown projective area in a cell is 0–10%
Shrub	Low woodland (grass, bamboo grass and shrub covered)
Sparse woodland	Crown projective area in a cell is 10–70%
Dense woodland	Crown projective area in a cell is 71% and over

dense woodlands are more significant, whereas sparse woodlands, shrubs, and grasslands have decreased. Such homogeneity of woodlands was the result of neglect of satoyama landscapes as a bioresource and, hence, lack of management, the other main cause of satoyama landscape transformation after the war. The dense woodland in the center of this area became the first national government park called *Musashi Kyuryou Shinrin Koen* (Musashi Hills Woodland Park) in 1974, 4 years after the photo was taken. It is fair to consider that, in 1970, the area was under preparation for the park.

Figure 3.18c shows the land cover of the area in 1995. Dense woodland is even more significant, except for golf courses and public open spaces (grassed). A significant decrease in paddy fields cannot be identified, although the tendency was for paddy fields to be transformed into dry fields and building sites.

Figure 3.19 illustrates the proportion and transition of land use in the diagrams, based on the results of Figure 3.18. In 1947, dry fields occupied the greatest area, followed by paddy fields. What is called woodland now, which is dense woodland, occupied only 15% of the entire area, approximately the same proportion as

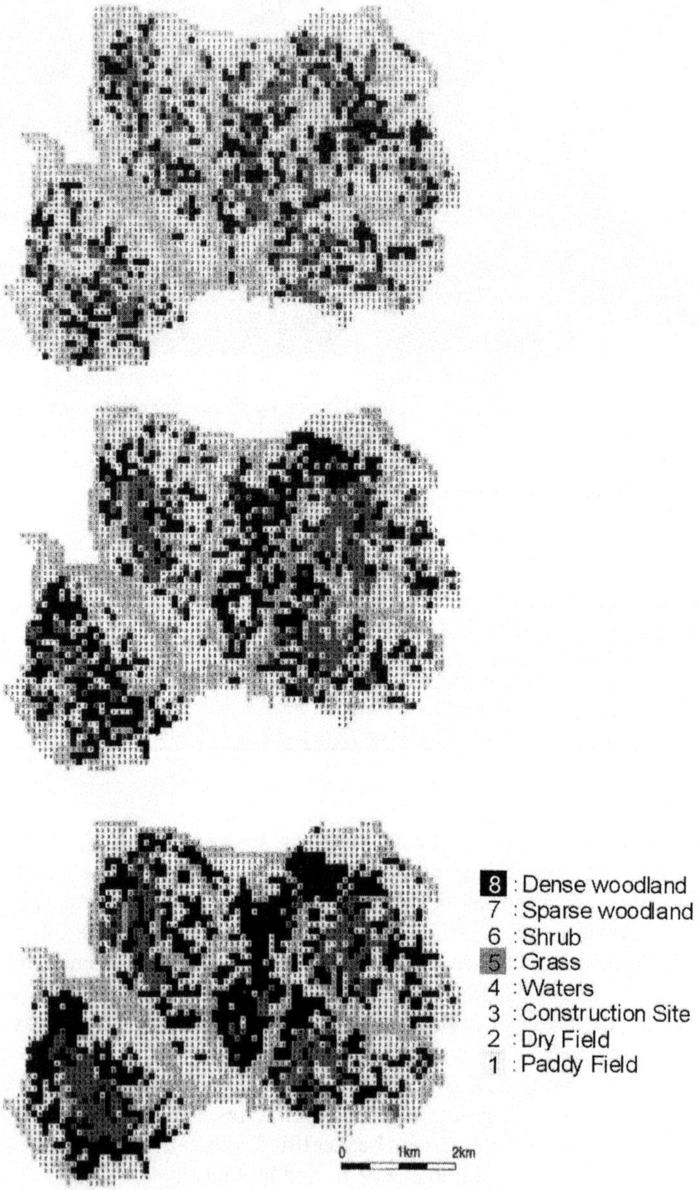

Fig. 3.18. Land-use maps of North Hiki Hills in 1947 (top), 1970 (middle), and 1995 (bottom)

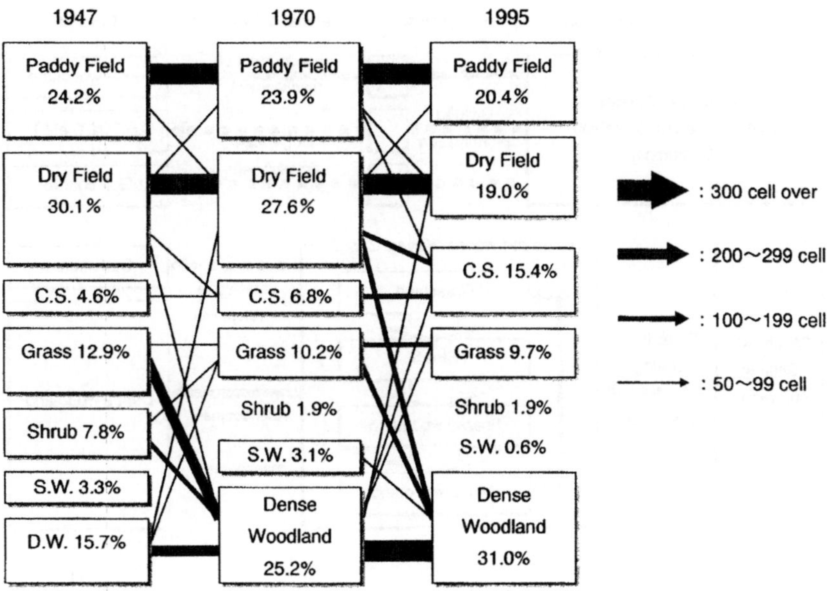

Fig. 3.19. Changing patterns of land use in the North Hiki Hills from 1947 to 1970 and 1995. C.S., construction site; S.W., sparse woodland; D.W., dense woodland

grasslands. Interviews with local farmers revealed that many of these grasslands occurred after *P. densiflora* or *Q. acutissima* and *Q. serrata* woodlands were felled. In fact, 10 trees ha^{-1} of mature *P. densiflora* were identified in this area. These were *tanematsu* (seed source *Pinus*). During the war, *P. densiflora* was not only felled for charcoal, but also for building material and *Pinus* oil. In 1943, the trees were supplied to the military as building materials. After the war, the new inheritance tax system promoted continuous felling of *P. densiflora*. Consequently, the remaining *P. densiflora* woodlands were significantly depleted during this time.

Pinus densiflora woodlands were greatly damaged by pine wilt decease after 1970, and many disappeared. Damaged *P. densiflora* woodlands transformed to *Q. serrata* woodlands as *Q. serrata* trees took over and grew, then partly succeeded into broad-leaved evergreen forests of *Q. myrsinaefolia* and to *Q. glauca*, which is the potential natural vegetation of the Hiki Hills. As a result, shrubs and sparse woodlands decreased sharply. These land covers can hardly be identified in 1995. Paddy fields decreased slightly between 1970 and 1995, but the decrease is not significant enough to be identified as a general trend. On the other hand, building sites and dense woodlands increased. Building sites doubled between 1970 and 1995, reflecting the wave of urbanization that washed across the area. Dense woodlands also doubled between 1947 and 1995.

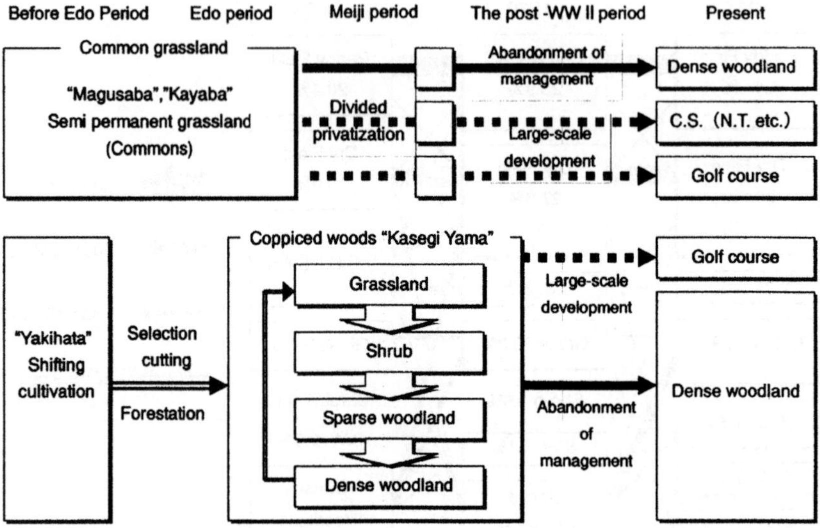

Fig. 3.20. Conceptual diagram of the changing patterns of satoyama landscapes since the Edo period. WWII, World War II; C.S., construction site; N.T., new town

From the figures, the following trends can be identified. Land use transformation was more stable between 1970 and 1995 than between 1947 and 1970. Between 1947 and 1970, transformation from grasslands to dense woodlands, and shrubs to dense woodlands was significant. The increase in dense woodlands between 1947 and 1970 was mainly the result of grassland and shrub transformation.

Between 1970 and 1995, transformation from dry fields to building sites, dry fields to dense woodlands, and grasslands to dense woodlands were significant. Dry fields were mainly transformed into building sites and dense woodlands, and dense woodlands were mainly transformed from dry fields and grasslands. In the Hiki Hills, as well as in other areas, the fuel revolution (from charcoal to oil and electricity) and the fertilizer revolution (from organic fertilizer to chemical fertilizer) occurred after the war. Satoyama landscapes lost their practical value and gradually became neglected. In addition, from the 1960s, farmers started side-line businesses and, after the 1970s, the opening of the Musashi Hills Woodland Park and the extension of big investment industries to the area increased employment opportunities, which accelerated satoyama landscape neglect. The increase in dense woodlands can be seen as a barometer of satoyama landscape neglect.

Grasslands did not show a significant change in proportion, but there was a dynamic transition from grasslands to dense woodlands, and vice versa, between 1970 and 1995. Transition from grasslands to dense woodlands was a consequence of coppice woodland becoming woodland. On the other hand, the transition from dense woodlands to grasslands was mainly the result of the development of golf courses. Thus, what appears to be the same grassland has totally different qualities.

From the analysis, it can be seen that satoyama landscapes underwent a dynamic transition through the pre- and post-war period, as illustrated in Figure 3.20. Detailed analysis reveals that there were large-scale causes, largely land use changes, and also periodic small-scale causes, largely due to the abandonment of the traditional land use and management, responsible for these transitions.

It is only fairly recently that a prototype of 'satoyama landscape = satoyama landscape woodland' has become popular. Previously, not only woodlands, but also spaces, such as grasslands and shrubs, were considered to be important components of satoyama landscapes. The mosaic of diverse types of spaces that existed as a result of management have turned into homogeneous dense woodlands as a result of the neglect of satoyama landscapes after World War II. This is something to consider when dealing with satoyama landscape conservation in the future.

In recent years, people have tended to focus on the woodland portion of satoyama landscapes. Satoyama landscape conservation should also include grasslands and shrubs, which are more at risk of disappearing, based on the recognition that satoyama landscapes are landscapes that have always been in dynamic transition. Only then can the identity and diverse ecology of satoyama landscapes be conserved.

4 Biological Diversity in Satoyama Landscapes

4.1 Conserving Biological Diversity

N. Kuramoto and Y. Sonoda

4.1.1 Diversity of Endangered Species

Satoyama landscapes have been identified as the primary habitat for several endangered plant species (Wagakuni niokeru hogojou juuyou-na shokubutushu oyobi gunnraku nikannsuru kenkyuuiinkai shu bunkakai 1989). Conservation and management of secondary nature areas has been emphasized as being necessary for wildlife conservation in Japan. The majority of the 351 plant species that are considered important in nature conservation inhabit secondary woodlands, while only 176 species are in climax forests (Table 4.1; Red Data Book Kinki Kenkyukai 1995).

This section will discuss the threat to species diversity in satoyama landscapes and consider options for conservation, with a special focus on plant and bird species. In general, people seem to hold one of two opinions about satoyama landscape conservation: they need active management; or they should be left unmanaged and allowed to succeed over time. The former opinion tends to be that of plant conservationists, whereas the latter is that of bird conservationists. This is one of the reasons why plant and bird species are the focus of this discussion.

4.1.2 Plant Species Diversity

Coppice woodlands are a typical vegetation cover in satoyama landscapes. These are, however, decreasing in quality due to loss of area, fragmentation, isolation, and change of management scheme. Nakashizuka and Iida (1996) analyzed the size and fragmentation of coppice woodlands in five Kanto areas using aerial photographs. The results indicated a great decrease in the amount of woodland adjacent to urban areas. Patches of woodland that are remote from urban development have also decreased, although the total area increased slightly in some areas (Table 4.2). There is a clear relationship between woodland area and species diversity (Figure 4.1) in logarithmic scale. Figure 4.1 indicates that, in order to conserve coppice woodlands that host twice as many species, fourfold more wooded area and sixfold more understory area will be required (Nakashizuka and Iida 1996). In general, the smaller

Table 4.1 Habitat of important plants in the Kinki district (After Red Data Book Kinki Kenkyukai 1995)

Habitat type		Fresh water wetland					
Habitat	Sum total	Water	Eutrophic marsh	Oligotrophic marsh	Field	Paddy field	Total of type
Gamopetalae	202	9	17	15	13	6	47
Schizopetalae	238	10	12	14	9	3	41
Monocots	277	31	35	33	4	9	86
Gymnosperms	2	0	0	0	0	0	0
Pteridophytes	143	6	3	7	1	6	11
Total	862	56	67	69	27	24	185

Habitat type	Sea				Grassland			
Habitat	Salt marsh	Sandy shore	Sea-shore	Total of type	Yama grass-land[1)]	Kaya grass-land[2)]	Sato grass-land[3)]	Total of type
Gamopetalae	4	8	8	19	18	14	22	41
Schizopetalae	4	8	15	27	19	10	29	52
Monocots	7	9	9	23	22	2	19	39
Gymnosperms	0	0	0	0	0	0	0	0
Pteridophytes	0	1	5	6	3	0	3	5
Total	15	26	37	75	62	26	73	137

Habitat type	Rock ground			Forest and woodland				
Habitat	Dry river bed	Rocks	Total of type	Second woodland	Evergreen forest	Total of type	Others	Specification is impossible
Gamopetalae	2	32	34	73	19	84	3	3
Schizopetalae	2	44	46	113	29	121	3	0
Monocots	1	39	40	79	67	116	4	11
Gymnosperms	0	1	1	0	2	2	0	0
Pteridophytes	0	48	48	86	59	106	4	6
Total	5	164	169	351	176	429	14	20

The satoyama has many important plants under protection.
[1)]The area except for wetlands and Kaya grassland among grassland in the mountains distant from the village.
[2)]Grassland like Japanese pampas grass, *Miscanthus sinensis* Anderss, among grassland near the village.
[3)]Grassland around crop land near the village.

Table 4.2 Areas and fragmentation of coppice woodlands

	Tokorozawa	Kasama	Iwai	Mito	Hachioji
Elevation(m)	40–70	70–190	10–20	10–40	100–300
Proportion of woodland land (%)					
1961	28.3	46.2	26.3	20.8	25.7
Recent years	20.8	48.2	22.0	16.7	15.8
Woodland stand density (per km^2)					
1961	5.0	6.9	11.0	9.4	4.7
Recent years	5.7	9.4	10.2	8.7	5.7
Mean area of woodland stands (ha)					
1961	6.5	10.7	2.5	2.3	6.8
Recent years	3.8	7.5	2.2	2.1	3.3
Mean minimum distance between stands (m)					
1961	35.0	27.0	41.0	42.0	50.0
Recent years	47.0	35.0	32.0	75.0	71.0

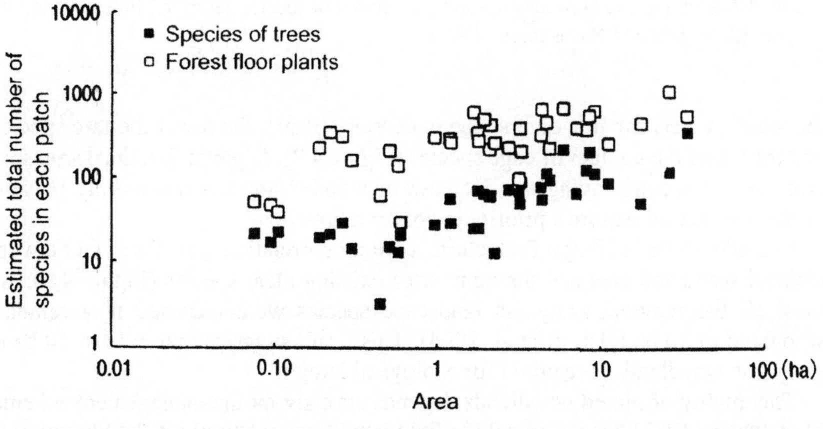

Fig.4.1. Relationship between area and the estimated number of species of trees (height > 2 m) and understory plants (height < 2 m). (After Iida and Nakashizuka 1995)

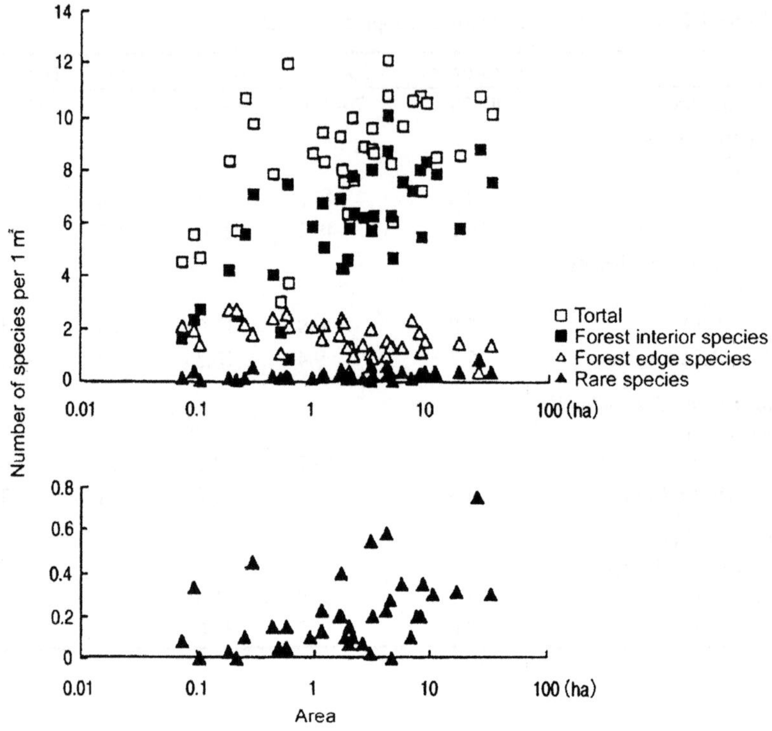

Fig. 4.2. Relationship between area and the number of species per m² of different habitat types (After Iida and Nakashizuka 1995)

the wooded area, the less diverse the understory plants, the fewer the rare species, and the larger the number of edge species (Figure 4.2). In particular, the decrease of rare species accompanying the decrease of wooded areas is noteworthy because those rare species require a priority in conservation.

In a new town in Hyogo Prefecture, a strong correlation was observed between isolated woodland area and the number of existing plant species (Figure 4.3). Almost all the regional satoyama landscape species were recorded in a remnant woodland of 10 ha (Hattori et al. 1994). This result suggests that at least 10 ha of satoyama woodlands is required for ecological integrity.

The quality of mixed woodlands depends strongly on the management scheme. Shigematsu (1985) has discussed the light conditions required for the blooming of azaleas and Yabu (1985) has discussed the conditions required for the blooming of *Erythronium japonicum*. The light conditions required for blooming were at least 30% for *Rhododendron reticulatum*; and at least 20% for *Rhododendron karmpferi*.

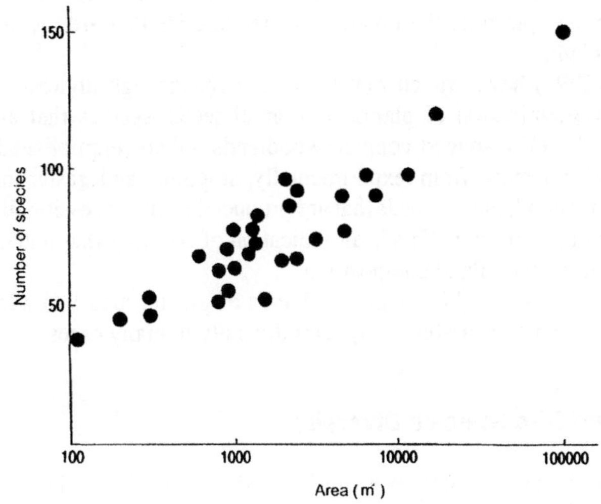

Fig. 4.3. Area of fragmented woodland and the number of plant species in the new town of Hyogo Prefecture (Hattori et al. 1994)

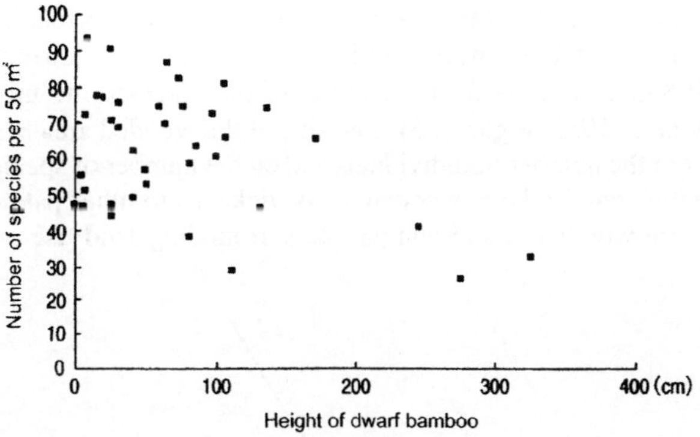

Fig. 4.4. Number of species of understory plants per unit area and the height of *Pleioblastus chino*

However, even 40% was not enough for *Rhododendron macrosepalum.* It is said that 40%–50% is desirable in springtime for *E. japonicum.*

Neglected coppice woodlands tend to become overgrown with thick *Pleioblastus chino* shrubs, resulting in a dark understory. According to Nakashizuka and Iida (1996), there is a negative correlation between the height of the bushes and the number of understory species per area (Figure 4.4.). There is definitely lower diversity underneath a thicker shrub community.

Hattori et al. (1995) have proposed that tall trees be maintained in satoyama landscapes. Removal of shrub species, while simultaneously maintaining existing tall trees, such as *Pinus densiflora*, coincided in a remarkable increase in the overall number of species (Yamazaki et al. 2000). Removal of the understory in *P. densiflora* secondary woodlands also results in an increase in the number of species

(Yamase 1998). However, too much disturbance may result in *Miscanthus sinensis* fields, or wasteland, or may promote the invasion of tree species that are incompatible with *Pinus* woodlands.

Tsuji and Hoshino (1992) have argued that litter removal through understory management allows the germination of plants with small seeds, species that are important components of well-managed coppice woodlands. Ohno (unpublished) has also reported similar results from experimentally stopping and restarting management. On the other hand, bigger seeds that are produced from successionally mature species can germinate in litter. This is an indication of one way that understory management can affect woodland composition.

In summary, species diversity is higher in woodlands of greater area and adequate vegetation management is desirable for species diversity in many cases.

4.1.3 Habitat Area and Bird Species Diversity

Woodlands in satoyama landscape have been influenced recently by residential development and road realignments. The resultant decrease in area, fragmentation, and isolation are adversely affecting the habitat for bird species as well.

In general, the positive correlation between the size of the habitat area and the number of inhabiting species is called species–area correlation. Such correlations are reported for the number of bird species and woodland areas for different woodland types (Yui 1988), as well as for the number of bird species and urban greenery areas (Higuchi et al. 1982; Figure 4.5). The size of the wooded area has a considerable effect on both the number of individuals and on the number of species. Abundant bird communities require large wooded areas, linkages to other patches of woodland and to a core woodland, and compatible surrounding land use or a

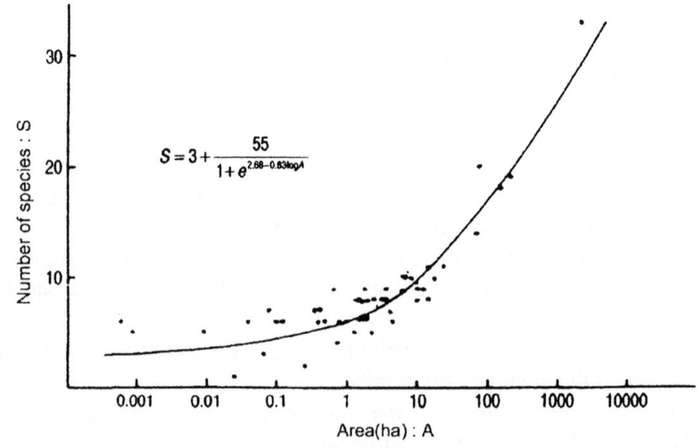

Fig. 4.5. Relationship between woodland areas and the number of species (Higuchi et al. 1982)

buffer zone (Ichinose and Katoh 1994). Bird species are also influenced by distances from other local populations (Natuhara and Imai 1999). Unlike plants, birds have a greater traveling ability, so that habitats where local extinctions occur can be repopulated from nearby or connected populations (Natsuhara 2000).

4.1.4 Vegetation Structure and Bird Species Diversity

Overmanaged woodlands can end up with a very simple woodland structure that will have low species diversity. It has been reported that the complexity of the vegetation structure (Figure 4.6) and woodland mosaic (Figure 4.7) are important factors for bird species diversity and high population density (Yui 1988; Yui and Ishii 1994).

Recent studies have started to focus on factors that impact bird species diversity in relation to vegetation structure and species composition in different types of woodlands. Data are being gathered to develop vegetation-management schemes for high bird species diversity.

In urban park woodlands, which are under greater human influence, deciduous woodlands are more important for bird habitat. In addition, dispersion of small coniferous patches and, in the case of no understory community, a mix of single- and multistem trees are important (Hayama 1994). Deciduous woodlands with little or no understory often resemble grasslands, which are preferred by grassland and woodland edge species such as *Emberiza cioides*, *Emberiza spodocephala* and *Lanius bucephalus*. A dense understory promotes bird species diversity (Hayama et al. 1996). In isolated woodlands, the complexity of the vegetation structure, espe-

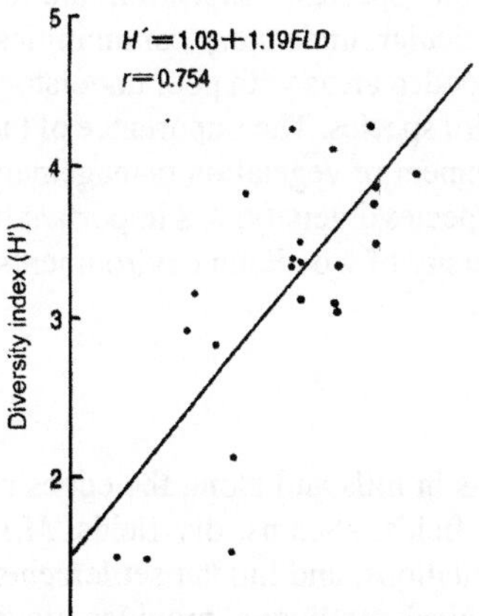

Fig. 4.6. Relationship between bird species diversity and woodland layer diversity (Yui 1988)

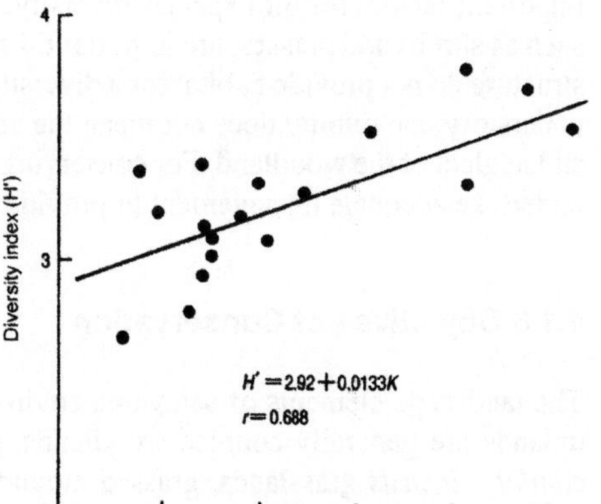

Fig. 4.7. Relationship between bird species diversity and the woodland mosaic (Yui and Ishii 1994)

cially the complexity of shrubs and herbaceous communities, plus dead trees, can positively affect species diversity (Ichinose and Katoh 1996). Some species, such as *Cettia diphone* and *E. spodocephala*, require dead trees for nesting. In upland parks, bird species composition varies for each different woodland type. The possibilities exist that increasing the grass cover area positively affects bird species that prefer scattered woodlands and that, also, greater shrub cover in breeding seasons results in greater species diversity.

4.1.5 Effect of Vegetation Management on Bird Species Diversity

In traditional managed satoyama landscapes, the woodland understory was managed regularly. Thick bushes were moved away and fallen leaves were cleared to allow easier access for charcoal production. However, it is clear that this type of satoyama landscape management is not necessarily desirable for conserving bird species diversity.

Some woodland bird species, such as *Cettia squameiceps* and *Zosterps japonicus*, prefer shrubs, grasslands with coniferous trees and shrubs, and secondary woodlands that are in the progress of natural succession into evergreen forests. Secondary woodlands with little understory community and pine woodlands with extreme management do not attract bird species that prefer sparse woodlands and grasslands (Ichinose and Katoh 1999). Development of *Pleioblastus chino* shrub communities may result in increases in the number of species, but the increase is limited until the proportion of the shrub cover is at least 40% (Katoh 1996). Understory occupation by shrubs does not enhance the diversity of bird species (Ichinose and Katoh 1998).

Thus, woodland area, complexity of structure, and species composition are the important factors for bird species diversity. In particular, understory communities, such as shrubs and grasses, are important. Large wooded areas with poor understory structure do not provide habitat for a diversity of bird species. The importance of the understory community does not mean the abandonment of vegetation management and neglect of the woodland. For conserving bird species diversity, it is important to undertake adequate management to provide a diversity of woodland environments.

4.1.6 Objectives of Conservation

The landscape elements of satoyama environments in hills and along the edges of uplands are generally coppice woodlands, paddy fields, streams, dry fields, *Miscanthus sinensis* grasslands, grassed mounds, plantations, and human settlements. The mosaic of these elements, reflecting geographical conditions, provides biodiversity in satoyama landscapes and directly affects wildlife ecology (Moriyama 1997).

The same theory can be applied to satoyama landscape conservation. The biodiversity of the satoyama landscape can be maintained by temporal and spatial mosaics of woodlands in different stages of succession. It is more meaningful to conserve this mosaic, especially in satoyama landscapes in larger areas. On the other hand, isolated smaller woodlands in urban areas are not large enough to maintain species diversity of all guilds. It is more suitable and feasible to decide the best guild to conserve within the surrounding context.

Coppice woodland management, which once was a labor for life by farmers, is transforming into labor for biodiversity conservation. It is desirable to find the procedure that best meets the needs of biodiversity conservation while respecting traditional methods. Management systems need to be developed for each satoyama landscape with appropriate goals and objectives, and sequences of monitoring, evaluation and feedback.

4.2 Species Diversity in Satoyama Landscapes

I. Washitani

4.2.1 Satoyama Landscapes and the Dynamic Bio-Interrelationship

Satoyama landscapes are mosaics of various natural environments. Diverse species survive as a result of less interspecies competition due to human intervention. The diversity of plant species provides habitat and food for a wide range of faunal species. Some species help the propagation of plant species by dispersing seeds and pollen and are, therefore, indispensable.

In satoyama landscapes, where patches of the same types of habitats are distributed in mosaics, the conditions of meta-populations of plant communities in woodlands and grasslands largely depend on the condition of faunal species that share the habitat (Washitani 2001a). In other words, mosaics of habitats in satoyama landscapes affect not only relationships between fauna and flora, but also gene flow and individual movement. This affects the spatial distribution and its genetic mutation. Satoyama landscapes offer such dynamic ecologic phenomena as well.

4.2.2 Herbivorous Insects in Satoyama Landscapes

Plant diversity is the foundation for biodiversity. From each plant species, an intricate food web is developed in satoyama ecosystem. It is not easy to reveal the entire picture of this food web. So far, only a small part of the web is understood.

Herbivorous species can be divided into two groups: generalists that consume several plant species; and specialists that consume only certain plant species. Herbivorous mammals and birds tend to be generalists. They consume a number of plant species and also consume large amounts because they themselves are relatively large. In contrast, there are many specialists among insects. The specialists cannot maintain their population unless they can be assured of a constant and abundant supply of consumable plants. Thus, it is advantageous for them to choose common plant species. It is for this reason that many insects in woodlands specialize in the consumption of woody plants. Although there will be less risk of a lack of food by specializing on common plants, common plants tend to develop more advanced protection against predators, which requires additional cost on the predator's side. The specialized predators are at an advantage because they are equipped to overcome the protection. The merit of specializing in minor plant species is that there is less competition, but there is the difficulty of ensuring enough supply for maintaining the population.

An order of insects that includes many specialist butterfly and moth species is Lepidoptera. According to Kato (2001), there are at least 85 species of Lepidoptera that consume *Quercus serrata*, which dominates coppice woodlands. Lycaenidae are popular for their beautiful, colorful appearance. Among them, there are a few specialists dependent on deciduous oak trees and also quite a few specialists dependent on plant species in coppice woodlands. In this sense, the Lycaenidae is one of the common butterfly species of coppice woodlands. The recent decline of coppice woodlands is said to have had negative effects on Lycaenidae.

The decline of coppice woodlands also negatively affects *Favonius orientalis*, which is a specialist dependent on *Quercus acutissima*, and *Japonica saepestriata*, which is a specialist dependent on *Q. acutissima* and *Quercus variabilis*. The only species that can be readily found in mixed woodlands around human habitation is *Antigius attilia*.

The numbers of *Neozephyrus taxila* are decreasing, accompanying a decrease in *Alnus japonica*, which is a wood species that grows at the waters edge in satoyama landscapes. The numbers of *Chrysozephyrus smaragdinus*, which is a specialist dependent on *Prunus jamasakura*, are also decreasing due to coppice woodland decline.

One of the symbolic butterfly species in satoyama landscapes is *Sasakia charonda* (Figure 4.8). Their larvae consume *Celtis sinensis*, one of the component woody species of coppice woodlands. The young live on *C. sinensis* and the adults live on the sap of *Q. acutissima*. Because sap is secreted only from younger trees, it is necessary that the trees are periodically cut down and regenerated for *S. charonda* to prosper. Conserved mature mixed woodlands without management do not provide an appropriate habitat for *S. charonda*.

Although they are not specialists, *Oryctes rhinoceros* also require managed mixed woodlands because their young grow in decayed deciduous trees and the adults suck on sap.

There are many butterfly species in danger of extinction from the decrease or abandonment of managed grasslands. The outstanding example is the decline of *Shijimiaeoides divinis*, which consumes *Sophora flavescens*, a species of the legume family. *Sophora flavescens* is left behind by grazing cattle, so pastures generally have many of these plants growing, thus assuring a desirable habitat for *S. divinis*. However, pastures have recently been abandoned and have been turning into shrublands. Thus, *S. flavescens* has been disappearing, along with *S. divinis*.

Similarly, many butterfly species are disappearing as grasslands decline due to either development or management neglect. Examples include *Leptidea amurensis*,

Fig. 4.8. A butterfly in coppice woodland

which consumes *Vicia amoena* and the adults of which feed on the nectar of *Viola grypoceras*, and *Maculinea teleius kazamoto*, which consumes *Sanguisorba officinalis* and parasites in anthills eating *Myrmica kotokui* as larva. The numbers of both species are said to be greatly decreasing due to the decline of *M. sinensis* grasslands.

For specialist insects, however, vegetation changes mean the loss of their habitat.

4.2.3 Mutualism in Satoyama Landscapes

The most effective and important pollinators of flowers are Bombinae, such as *Bombus ardens* and *Bombus diversus*. Like honey bees, Bombinae organize a colony and nest underground, hive in trees or crevice in artificial objects. However, as opposed to honey bees, Bombinae colonies last only 1 year (Washitani et al. 1989).

The primary food of Bombinae is flowers. They gain carbohydrates from the nectar and protein from the pollen. The relationship between Bombinae and flowers is that each individual species specializes in a certain species of flower. This is an ideal situation for plants, because their pollen will be transported to other plants of the same species. In developing such partnerships, plants evolved several devices to deter other insect species. For example, some species of plant have developed an elongated flower so that only bumble bees with long tongues can reach their nectar, or they have flowers facing downward so that only bumble bees can grab on to the petals upside down and reach the nectar. These specially evolved flowers for bumble bee pollination are known as bumble bee-pollinated flowers.

In satoyama landscapes, bumble bee-pollinated flowers stagger their flowering times in order to avoid interspecies competition. Continuous provision of flowers is an essential condition for Bombinae habitat. On the other hand, Bombinae is a crucial biological factor for bumble bee-pollinated flowers and plays an important role in seed production. Through intensive observation of *Primula sieboldii*, there are several recorded cases of the plants not producing enough seeds when bumble bees were not involved in the pollination process.

A colony of Bombinae is formed with a queen after hibernation and then collapses, leaving a number of new queens. The collapse of *B. ardens* colonies occurs in early summer, whereas that of *B. diversus* colonies occurs in autumn. The new queen awakes after hibernation, lays eggs, and raises worker bees. Until the worker bees grow and start working, the queen collects pollens herself. *Primula sieboldii* have chosen queens of *B. diversus* as partners (Washitani 1998). They have adjusted their flowering season to the period of the queen bee's pollen collection and have also evolved their flower form to adapt to the queen bee's form, which is a larger body and long tongue.

In springtime, white flowers with short collora facing down occur on several woody satoyama landscape species, such as *Styrax japonica*. *Bombus ardens*, which do not have a long tongue, is the partner for these plants. *Bombus ardens* also partners with *Wisteria floribunda*, which lives on the woodland edge and covers itself with purple flowers in springtime.

Bombus diversus visits any kind of flower and transports pollen. It is queen bees in spring and worker bees in early summer and autumn that contribute to the transport of pollen. On the other hand, *B. ardens* acts during a relatively short period of time from early spring until before the Japanese rainy season in June. *Bombus ardens* contributes to the pollination of many woody species that flower in the spring.

The major nesting sites of Bombinae are abandoned nests of small animals, such as rats. It is probable that not only the destruction of woodlands, but also a lack of maintenance is leading to a break in the continuous flower provision of the plants in satoyama landscapes and adversely affecting the survival of Bombinae. Research to prove this theory, however, has not yet been undertaken

4.2.4 Seed Dispersal in Satoyama Woodlands

Plant species have limited dispersal ability. Seed dispersal over long distances often depends on wind and transportation by animals. Seed dispersal by bird species plays a significant role in satoyama landscapes, where woodlands typically develop in rather smaller patches. On the other hand, ant species are said to play an important role for the seed dispersal of understory plants in coppice woodlands. Ants and birds have substantial differences in terms of the distances they travel. Birds typically travel long distances and distribute seeds of edge species and shrubs of coppice woodlands, whereas ants travel relatively short distances and contribute to the distribution of understory plants.

Bird species contribute to seed dispersal in several ways. Here, we focus on Endozoochorous seed dispersal, which is seed dispersal by birds as a result of fruit consumption. Adaptation to Endozoochorous seed dispersal is so common, that it is found in half of all angiosperm plants. The importance of the system in the dynamism of woodlands has been well recognized for a long time and has even been applied in methods for the reestablishment of vegetation. For example, during the Edo period, Banzan Kumazawa developed and applied a method for the restoration of mountains utilizing seed dispersal by bird species (Kominami 1993). It was done simply by scattering bird food in areas that were to be vegetated. Birds would come and eat the food and drop seeds with their feces. The seeds grew to be a woodland without much management. This is truly a natural method of vegetating areas.

In satoyama landscapes in the Kanto area, there are several possible candidates for seed dispersal agents. There is a great difference in dispersion distance and patterns between migrating and non-migrating birds, but they both play important roles in dispersing seeds of Japanese plants with juicy fruits.

Dispersal distance depends on how far the bird travels while the seed is in its gut. A few hundred meters or more is common (Fukui 1993). It is also expected that transportation from one woodland patch to another is possible.

Among the understory plant species in coppice woodlands, especially those termed spring ephemeral plants, there are several that have ants as seed dispersal agents. It is easy to distinguish these seeds because they are small enough for ants to carry and have eraiozome to attract the ants. In coppice woodlands, many plants depend on ants for seed dispersal. The dispersal distances covered by ants are not great. For example, some ants carry seeds only approximately a few decimeters, which is far shorter than the dispersion distance of seeds popping out from pods (average 2 m, maximum 5 m; Suganuma et al., unpublished).

Ants carry seeds with eraiozome into their nests. They remove the eraiozome to be used as food, then remove the seeds from their nest and dump them. Thus, seeds are dispersed fairly close to ant hills. Seed dispersal by ants is ideal for plants that prefer the same environment where ants nest, such as vegetation gap areas or woodland edges.

Plants that depend on seed dispersal by birds can expect recovery, even when they have disappeared from a woodland, as long as there are the same types of woodland nearby. Birds can reseed the area. On the other hand, seed dispersal by birds can be the cause of some problems in coppice woodland conservation. The importation of evergreen woody species and introduced horticultural species may transform the species composition of woodlands into evergreen woodlands. This type of woodland creates an environment that discourages the typical understory plants of satoyama landscapes to grow. Isolated woodlands in urban areas are more likely to confront this situation.

On the other hand, plants that depend on seed dispersal by ants are difficult to recover once they have disappeared from an area. This is even more so in isolated woodlands.

4.3 Wetland Environments and Biodiversity in the Hills

Y. Morimoto

4.3.1 The Wildlife of Shinodayama

Dog walkers, joggers, plant-, insect- and bird-lovers, and children all use the area of Shinodayama as a leisure facility. On the whole, however, the area lacks facilities. For this reason, it is a particularly beautiful patch of countryside, with its winding, unkempt paths. Unfortunately, the subtle folds in the landscape result in a lack of clear views, which has meant that illegal dumping has taken place largely unnoticed. Increasingly, the value of this resource has come to be recognized in other terms.

In Shinodayama, *Habenaria radiata* and *Pogonia japonica*, for example, have come under danger of extinction. Luckily, this area has provided a habitat for the world's smallest dragonfly and other rare species. Several decades ago, these species were common, whereas now their numbers have declined drastically. The preservation of the existence of these valuable species has now become a central objective of conservation in Shinodayama.

4.3.2 Endangered Species

Table 4.3 is a list of endangered species contained in the Red Data Book (Environment Agency 2000, Red Data Book Kinki Kenkyu 1995) that are found on Shinodayama. However, these species are not frequently found in this area. Already, only a very limited number can be found. Shinodayama is one of the largest clusters of wetlands in Osaka Prefecture. There is some question, however, of how long this wildlife can survive in the future. The Laboratory of Landscape Architecture and Conservation at Osaka Prefecture University and the Osaka Green Trust have surveyed Shinodayama and identified the current situation of wildlife there.

There is a considerable variety of species, from those that prefer a dry environment to those that prefer a wet environment. However, one thing these species have in common is that they all prefer non-woodland environments. The following investigates the details of these environments

Eulalia quadrinervis prefers grasslands and land prepared for development. This species prefers land that is very dry and nutrient poor, with plenty of sunshine. This species cannot live in virgin or thinned woodlands and is found on bare mountains tops. A similar grass species, namely *Eulalia speciosa*, is found on the hills near the Nagoya area, Kaisho Forest in Seto City (in Aichi Prefecture). This forest became famous as being the planned site for Expo 2005. The environmental impact assessment for the Expo area used *E. speciosa* as an indicator species for assessing the impact of the development, because this species requires a particular environment. Table 4.3 contains another endangered species, namely *Platycodon grandiflorum*, which is also a plant usually found in grasslands and not in the woodland.

Table 4.3. Threatened plants in the Shinodayama Hills

Family	Species	Japanese red list category*	Red data book of Kinki district
Droseraceae	*Drosera peltata* var. *nipponica*	VU	X
	D. spathulata		X
Rosaceae	*Sanguisorba tenuifolia* var. *parviflora*		X
Asclepiadaceae	*Cynanchum nipponicum* var. *glabrum*		X
Alismataceae	*Sagittaria aginashi*	NT	X
Liliaceae	*Hosta longissima*		X
Poaceae	*Agrostis nipponensis*	NT	X
	Eulalia quadrinervis	VU	X
Cyperaceae	*Eleocharis fistulosa*	EN	X
	Scleria parvula		X
	S. rugosa		X
Orchidaceae	*Habenaria radiata*	VU	X
	Pogonia japonica	VU	X

*Ministry of the Environment.
EN, critically endangered; VU, vulnerable; NT, near thretened.

Table 4.3 also contains plants that prefer wet environments (e.g. *Habenaris radiata* and *Pogonia japonica*, which belong to the orchid family). These plants prefer small wetlands with nutrient poor, very clear spring water. In Shinodayama, you can find such wetlands with good spring water. In addition, these species require an environment with plenty of light. As well as members of the orchid family, these wetlands contain insectivorous plants, of which 10 types are found in the Osaka area and eight are found in the Shinodayama area.

One of those insectivorous plants, *Drosera peltata* var. *nipponica*, prefers an environment that is only wet after the rain but is usually dry, bare soil. In Shinodayama, you can find such patches on the ridges of the hills where *Pleioblastus chino* is also found. So, it is interesting that *P. chino* and *D. peltata* var. *nipponica* are creating a community on the upper parts of the hills.

Between the hills, there are small, shallow valleys, which are given over to the cultivation of small rice paddies. Today, most paddies are already abandoned. You can also find irrigation ponds in the valleys. Upstream of these, in the valley, you can see *Sanguisorba tenuifolia* var. *parviflora* communities, which prefer an environment with nutrient-rich water.

In summary, from the hydrologic perspective, you can find a whole variety of plant species that prefers a dry environment to a wet environment, which characterizes the area. Their commonality is that these species prefer a non-wooded environment. This non-wooded environment is quickly disappearing in Osaka Prefecture. According to the Osaka Green Trust, the total land area of these wetlands has decreased from 16.3 to 4.6 ha in the past 15 years.

4.3.3 Wetlands from Spring Water in the Osaka Stratification

The southern part of Osaka City, despite including the name mountain, has an elevation of only 70 m. These hills are characterized by Pliocene and Pleistocene sedimentary layers, formed when the area lay at the bottom of the sea before being raised to its current elevation. The rock layers are not very old, but are harder than soil. This layer surrounds the Osaka Plain, where you can find the hills today.

On these hills, you can find wetland created by spring water on the layers. These wetlands are small in size and found in unexpected places, such as the middle of the slope. This is because some of the layers are more permeable than others. The rainwater filters through a layer and then runs along an impermeable layer to come out at a different place. In terms of the climate zone, the Osaka area belongs to the Inland Sea; it is the area with the smallest amount of annual rainfall in Japan. This explains the preponderance of irrigation ponds. These irrigation ponds rely on spring water. The name of this area, *izumi*, means spring. Through this, we can understand how indispensable the springs were for the rice cultivation in the area.

The Shinodayama area is almost in the front line of nature conservation in Osaka. From ancient times, this area has been under intense human impact for rice cultivation or for harvesting firewood or leaf-litter as organic fertilizer.

4.3.4 Disappearing Nature and Protected Areas

The reason why this area has been protected is because it was used as a training zone by the Japan Self Defense Forces (JSDF). The surrounding area is totally developed, which took place in Osaka from the 1960s on a large scale. Because the hills are not very steep, they were relatively easy to develop. At first, Kori Danchi

Fig. 4.9. In 1969, there were mostly grasslands and barren lands (*left*). Plant succession has led to the development of woodland communities by 1998 (*right*)

was developed in the east and then the northern Senri Hills, which eventually became the Senri New Town, site for the Expo of 1970, were developed. The hills of Senboku, to the south of the Osaka Plain, have also been developed by the former Japanese Housing Corporation. Consequently, the area surrounding Shinodayama is mostly composed of large-scale housing developments. From a hazard prevention point of view, good drainage was necessary. The spring water wetlands came to be thought of as dangerous places, where mosquitoes and snakes breed. This led to people thinking that these places were not worth having anymore. Conservation was made difficult, if not impossible, through direct development, as well as people's attitude to nature. During the 1960s and 1970s, no one thought this could be important, because they imagined that this nature could be found anywhere. Today, the nearly unique environment of Shinodayama comprises 300 ha of grasslands, paddy, abandoned paddy, and pine woodlands with nutrient-poor soils (Figure 4.9).

The reason why such an extensive area of natural land is available is because of the army and the JSDF, which has regularly used and maintained the area for training since the Meiji period.

4.3.5 Deterioration of Wetlands and Woodland

Even though this whole land area has been preserved, the wetlands are now quickly disappearing or deteriorating. This is not because of the training or because of the harvest of trees, but because the whole land area has been isolated by urban development. This means that the ground water system has been changed. Another reason for the rapid disappearance or deterioration of wetlands, is because of abandonment, which has resulted in more and more of the area becoming wooded. As described above, because most of the plants prefer a bright environment with sunshine, this forestation reduces their habitats. Moreover, the more plant species that surround the wetlands, the more water is drawn off, drying the area. Even though the whole land area has been in use for training the army, some areas were included for farmers. These are mostly paddy fields. Farmers used the grass from the grasslands to make compost for fertilizer. Looking at the land use changes of

Table 4.4. Land cover type changes in the core area of Shinodayama (percentage per km^2)

Land cover type	1887	1968	1998
Woodland	1.6	16.9	42.5
Shrub land	0.0	7.3	6.5
Grassland	68.0	38.2	19.5
Barren land	13.6	0.0	0.0
Cultivated field	8.2	22.2	8.1
Residential land and roads	8.5	15.4	23.4

this area in a study area in the heart of Shinodayama, land use changes were identified using topographical maps and aerial photos. In 1887, almost 70% of the land area was abandoned *arechi* (rough ground) and was dominated by tall grasses; bare land comprised 10% and woodland 2% of the land area. By 1969, arechi had decreased by half and, by 1988, and it had decreased to less than 20%, whereas woodland had increased rapidly, now comprising more than 40% of the land area, with cultivated land being less than 10% (Table 4.4). As a result, the whole land use of Shinodayama has shifted from a bare or grass environment to a woodland.

4.3.6 The Osaka Sedimentary Layers and Biodiversity

In general, natural succession may be seen to be a good thing. In fact, when you think about the ability to capture CO_2, of course, woodland acts as a more effective carbon sink. The situation with biodiversity is not so simple.

The area of the Osaka Sedimentary Layers has a very fine textured soil. The soil quickly becomes firm and a great variety of tree species cannot be supported. Thus, pines in the area are very short and thin. In those places where urban development has taken place, there are serious problems with the trees because they won't grow so well in the poor soil. This kind of unsuitable soil is found when the layer has been under the sea because it contains a lot of organic matter. This causes deoxidization in the soil, which causes the sulfates in the sea water to be deoxidized and triggers the formation of pyrite layers in the soil. When these are excavated, they are re-oxidized in dehydrated form and this leads to the formation of sulfuric acid, making the soil too acidic and unsuitable for the growth of trees or plants.

Low precipitation and poor soils are the major characteristics determining biodiversity in the area. These two factors constrain woodland growth. This means that the trees in the area are not really good for making firewood or charcoal. Very acidic soil and varied conditions, spring water, and nutrient-poor wetlands have also contributed to the biodiversity in the area.

4.3.7 Conservation of Wetlands

Because the wetland is endangered, the Osaka Green Trust has started to conserve the wetlands on behalf of the Japanese Ministry of the Environment. The Osaka Green Trust started work on one of the remaining wetlands in the valley, which used to be rice paddies, but has been abandoned. Although this is strictly a spring wetland, springs are present in six locations in the area. This group has been working on restoring the wetlands in the area. They have worked on cutting the plants in the wetlands and excavating the roots of the bamboos in the area. The group has also restored the waterways, which are shown in Figure 4.10. The Osaka Green Trust has also restored a small spring on one of the slopes, as shown in Figure 4.10, and they have restored a small spring on one of the slopes surrounding the valley. They have transplanted several endangered species from another part of the

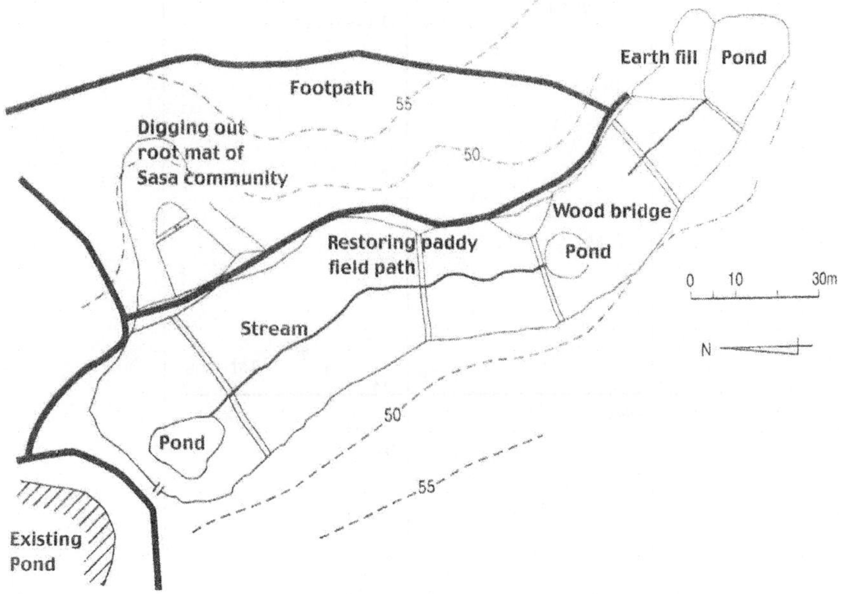

Fig. 4.10. Management techniques applied to a marsh in Shinodayama Hill by the Osaka Green Trust

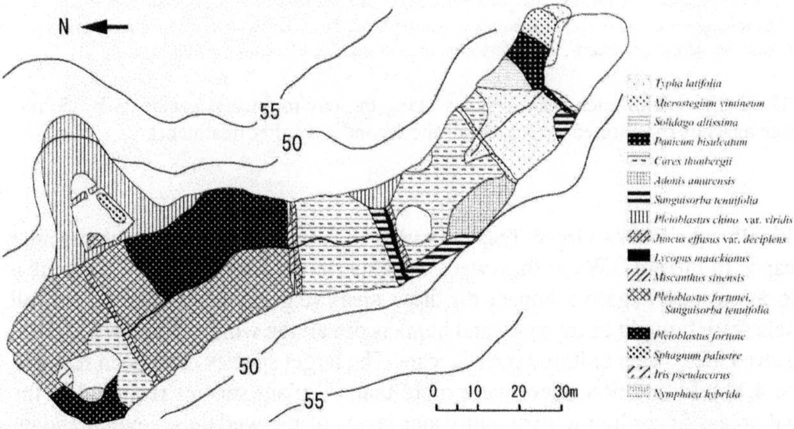

Fig. 4.11. Vegetation map of the marsh in Shinodayama Hill at the end of the second year after treatment

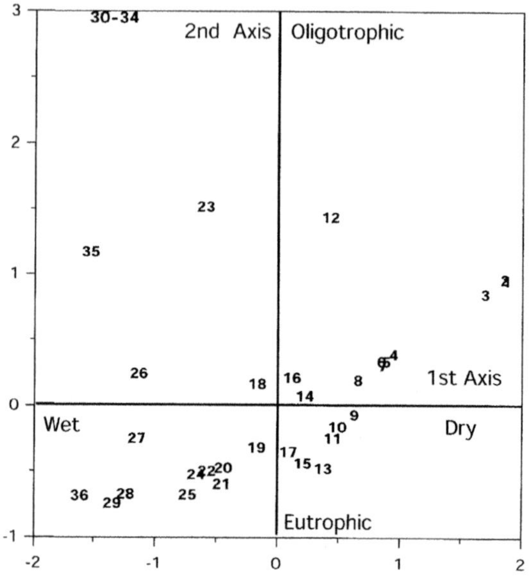

1, *Haloragis micrantha*; 2, *Hypericum laxum*; 3, *Centella asiatica*; 4, *Pleioblastus chino* var. *viridis*; 5, *Luzula capitata*; 6, *Millettia japonica*; 7, *Kummerowia striata*; 8, *Miscanthus sinensis*; 9, *Panicum bisulcatum*; 10, *Solidago altissima*; 11, *Microstegium vimineum* var. *polystachyum*; 12, *Cyperus pilosus*; 13, *Persicaria hydropiper*; 14, *Juncus effusus* var. *decipiens*; 15, *Arthraxon hispidus*; 16, *Bidens frondosa*; 17, *Rosa wichuraiana*; 18, *Eupatorium lindleyanum*; 19, *Persicaria pubescens*; 20, *Lycopus maackianus*; 21, *Murdannia keisak*; 22, *Sanguisorba tenuifolia*; 23, *Sium sisarum*; 24, *Persicaria nipponensis*; 25, *Typha latifolia*; 26, *Cyperus haspan*; 27, Pleio*blastus fortunei*; 28, *Carex maximowiczii*; 29, *Leersia japonica*; 30, *Rhynchospora chinensis*; 31, *Sphagnum palustre*; 32, *Sacciolepis indica*; 33, *Eleocharis congesta*; 34, *Mosla dianthera*; 35, *Triadenum japonicum*; 36, *Eleocharis wichurae*.

Fig. 4.12. Ordination of each plant species along the environmental gradients by correspondence analysis (performed in October of the second year after treatment)

hill. Finally, the Osaka Green Trust is trying to restore traditional or vernacular landscapes in the area. When they were working on the wetlands, they took great care to avoid any negative impact on these areas (e.g. no heavy machinery, all materials were brought in by hand, and humans did all the work themselves). Then, the restored area was monitored over 2 years. The target species have been restored (Figure 4.11). In addition, there were more than 70 plant species recorded in the restored areas. According to different water levels of the wetlands, several endangered species were also identified. According to the research conducted by Osaka Green Trust, the limiting factors for all these plant communities are the water level and nutrients in the water. Figure 4.12 shows representative plant species in these wetlands by showing the water level as the horizontal axis and the nutrient level as the vertical axis. When these lands are again abandoned, bamboo will intrude into the area again, which may dry the land and cause succession to take place. If the land is disturbed too much, then introduced species, such as *Bidens frondosa*, will intrude. So, optimum and constant maintenance is necessary.

At the valley bottom (former paddy fields), the water and soil are relatively nutrient rich. However, branch valleys contain spring water, places with nutrient-poor conditions, which provide good habitats for carnivorous plants; there has been a successful transplanting scheme for this. For these nutrient-poor wetlands, you may also need constant and appropriate maintenance. For other places, with no maintenance and where succession has been allowed to take place, it is clear that the endangered species will become extinct.

4.3.8 Keys to Conservation: Citizens and the Japan Self-Defense Forces (JSDF)

These days, people are more concerned with lands and the satoyama landscape. These movements are good on the one hand and bad on the other because they preserve the environment while allowing acts such as the stealing of endangered plant species. In March, 2000, Osaka Prefecture implemented policies with the aim of conserving the natural environment. Areas were designated as a protected area of wildlife species as a way of stopping the stealing of endangered species. However, it is doubtful whether this policy is working. In fact, there are a couple of local nonprofit organizations (NPOs) that have worked with nature conservancy. The relationship between the Osaka Prefectural Government and the local NPO is important in this respect. These NPO groups may not have enough information about nature and may bring in introduced species from abroad. To prevent such mishaps, a group of specialists should also be involved in such conservation works, to formulate the target of the work and to advise on what has to be done on site. Another indispensable element is the cooperation of the JSDF. Most of the spring water areas are found in the army training field. Also, disturbances made by training have had a positive impact on the ecology of the area, creating ideal conditions for some of the species. This irony is not only found in Shinodayama, but also in Ibaraki, where training fields were established for the United States Army but then became a park. In the same way, the woodland areas and sand dune habitats in these places have been maintained while limiting the access to and development of these sites.

In Japan, almost 70% of the land area is forest and woodland, while grassland comprises less than 1%. In Osaka Prefecture, Shinodayama is the only place where you can find grassland with the unique habitat necessary for the survival of certain plant species. It goes without saying that such areas need to be preserved.

At present, looking at the world, the United States Army Corps of Engineers has had the longest history in the world of conserving wetlands (Smith et al. 1995). Because they publish a manual for the protection or maintenance, such places (Environmental Laboratory, USACE 1987), when dealing with wetlands, the agreement of engineers is a necessary part. The JSDF does not have the same resources; nonetheless, in some cases, such as in Shinodayama, their cooperation and relations with them become a key issue.

4.4 Birds of Prey Living in Yatsuda and Satoyama

A. Azuma

4.4.1 Satoyama and Yatsuda

Paddy fields found in and around satoyama landscapes are mostly those on the mountain slopes as terraced paddies or in the valleys between hills and uplands, which are known as yatsuda. Kanto yatsudas will be described using the area of Tegan Marsh and Inba Marsh in Chiba Prefecture as an example. The area where you can find satoyama landscapes is mostly where you can find hills and uplands. Quite often, among these, there are small valleys known as yatsu. Because they are valleys, they are rich in springs, and this meant that little effort was needed for irrigation in ancient times. Thus, the yatsu determined the settlement patterns and cultivation in these areas. The rice paddies in these areas are called yatsuda. The local examples of the Tega Marsh and the Inba Marsh are explained in detail below. These marshes are located in Chiba Prefecture, where the land is flat. The average elevation is only 49 m. Consequently, this prefecture is known in Japan as the lowest. The area of Inba Marsh and Tega Marsh in the northern part is especially flat, consisting mostly of alluvial plains and lowlands. The landform can be divided into three categories: uplands; woodlands on terrace scarps; and paddy fields in the valley bottoms. On the uplands, the lands are mostly used for crop fields and settlements, being covered mostly by the Kanto loam layer. The woods on the edge of the uplands have mostly been used for charcoal or firewood and for fertilizer. Because they have lost their economic viability, these places are mostly abandoned. The typical vegetation of these woodlands includes the following: deciduous *Quercus* as secondary woodlands; *Chamaecyparis obtusa* and *Cryptomeria japonica* as planted species; and broad-leaved evergreen *Quercus* trees. This kind of landscape can be found not only Chiba, but also in Ibaraki, Tochigi, and other prefectures on the Kanto Plain.

4.4.2 Aquatic and Terrestrial Habitats Necessary for Wildlife

Yatsudas are places where you can find aquatic environments, for example as paddy fields, and as drainage and irrigation canals. The terrestrial environment of such areas is characterized by dikes or low embankments surrounding the fields. Surrounding the yatsudas, there are crop fields, irrigation ponds, and woodlands that are distributed according to the land form of the area. That kind of satoyama landscape provides an ideal terrestrial and aquatic habitat for animals such as frogs and amphibians. These are eaten as prey by raccoons, weasels, herons, snakes, and others (Ikeno 1993). So, frogs can be regarded as small animals that support the wildlife in the area (Osawa and Katsuno, 1997, 1998). Those frogs living in Japan quite often use paddies as places where they lay eggs. In Chiba, 10 different species of frog are found in total and eight of these species lay their eggs in the yatsuda.

Among those eight species of frog, the Wrinkled frog (*Rana rugosa*) and the Tokyo daruma pond frog (*Rana porosa*) are frogs that live on the land, but may not stray far from the yatsuda. On the other hand, there are frogs such as the Japanese brown frog (*Rana japonica*), Montane brown frog (*Rana ornativentris*), and Eastern-Japanese common toad (*Bufo japonicus*) that come to the yatsuda to lay eggs, but live in the surrounding woods or grasslands. Then, there is yet another group of frogs that lives on trees: the Japanese tree frog (*Hyla japonica*), Schlegel's green tree frog (*Rhacophorus schlegelii*), and the Forest green tree frog (*Rhacophorus arboreus*). Why is it that all these frogs prefer the yatsuda?

Ordinary paddies will have dried up during the winter and will have water only during the cultivating season. Yatsudas commonly contain water all year long. Some frog species lay their eggs in early spring, when other paddy fields are still dry and the yatsuda may be the only source of water. Also, the water in the yatsuda is spring water and, thus, warmer than water from melting snow. In addition, the size of the paddy is a factor: in general, each of the rice paddies is much smaller than those on the plains. This means that, quite often, the paddy will be surrounded by the woodland nearby and the distance to the edge embankments will be much smaller. This means that the distance for the frog to move about when matured and laying, or for the tadpoles to climb out of the water, is lower. Some of the toads also find this an advantage because they are not very active.

4.4.3 Yatsuda Changes Influencing the Frog Community

Yatsudas are indispensable for most of the frogs, but land readjustment has taken place to improve the agriculture. This has meant that paddy fields that used to have water all year long have had their drainage improved. So, they have no water in winter. Also, some of the paddies have been abandoned. Looking at the woodlands, changes have occurred in the vegetation because of abandonment. So, the environment is changing rapidly. How would the structure change as a result of farmland consolidation? The purpose is to increase yield and efficiency. To achieve this, the paddy has been dried using modern irrigation and drainage techniques. In addition, the size of the paddy has increased, and the drainage channels of bare earth have been replaced by concrete and straightened ducts. Irrigation, which used to depend on the springs, is now supplied with pipelines providing water from a distance. The roads will be paved to accommodate large machines. Such changes mean a serious impact on frog habitat. Our research group has studied 16 paddy field areas in the Tega Marsh and Inba Marsh areas and recorded the communities of *R. japonica* and *H. japonica* (Figure 4.13; Azuma and Takeuchi 1999). In this study, among the many factors that could have indicated changes in the yatsuda structure, the relationship between average density and several environmental indicators has been investigated. In particular, whether a modern drainage system exists, where water comes from (pipeline or spring) and differences in the design of the drainage canals have all been assessed. The population density of frogs was then surveyed. Where pipelines and canals with concrete are found, the population density of the *H. japonica* was significantly higher than in other places. Where land improvement

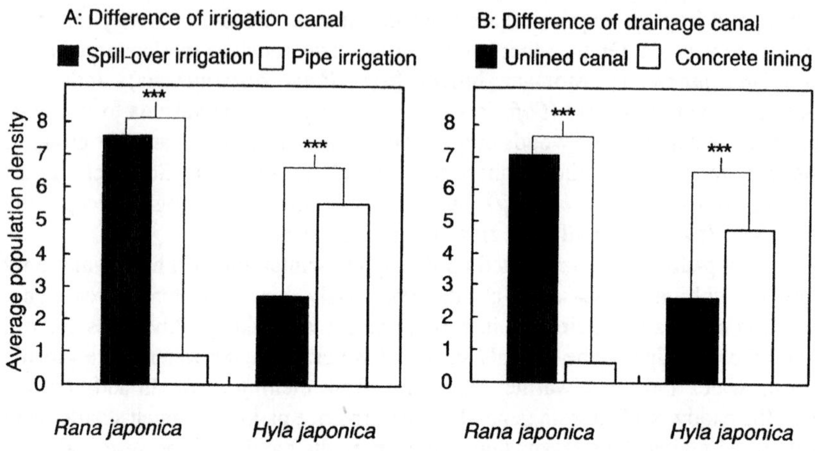

Fig. 4.13. Comparison of average population densities (per meter of levee) of two kinds of frog (*Rana japonica* and *Hyla japonica*) in different irrigation (*a*) and drainage (*b*) canals (After Azuma and Takeuchi 1999). $***P < 0.001$, Mann-Whitney *U*-test

cannot be found or where the impact of this is low (spring water remaining etc.), the population density of *R. porosa* is significantly higher than for other areas. This means that *R. japonica* can be found before the land is improved, whereas *H. japonica* can be found after land readjustments.

Another survey was conducted in the same area considering mature adult populations of *R. japonica*, *H. japonica* and *R. porosa*. Similar environmental factors were used as in the previous study. The rice paddies were classified into four categories according to differences observed in the three factors. Type 1 is without a drainage system and depends on spring water and bare soil drainage canals; type 2 has a drainage system, spring water supply and bare soil drainage canals; type 3 has a drainage system, water supplied by pipeline and bare soil drainage canals; and type 4 has a drainage system, water supplied by pipeline and concrete-lined drainage canals. Figure 4.14 shows the average number of animals found in each of the four types of areas.

With *R. japonica*, there was a big gap between types 2 and 3. With *R. porosa*, a similar gap was identified between types 2 and 3, but there was also a big gap between types 3 and 4. The reasons for this can be described in terms of life cycle and behavioral ecology. *Rana japonica*, which lays eggs in early February, needs water in winter. This means that a paddy without a drainage system is suitable, as is a paddy with drainage but with spring water (water all year round).

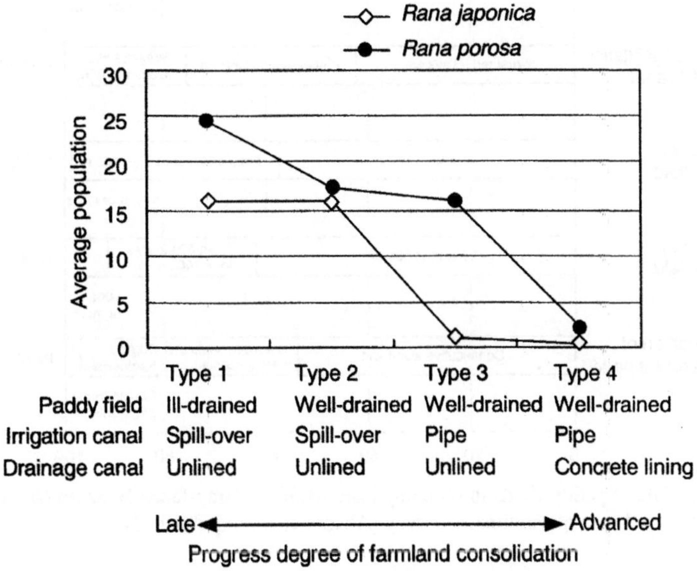

Fig. 4.14. Average population of two kinds of frog (*Rana japonica* and *Rana porosa porosa*) in four types of paddy fields with different degrees of farmland consolidation (After Azuma and Takeuchi 1999)

In contrast, *R. porosa* lay eggs after the rice has been planted (Fujioka and Lane 1997). Then, even after *R. porosa* become mature adults, they will remain around the paddy areas, so that the water shortage during winter does not affect them However, when the paddy is dried up after harvesting, another water body is needed to for these frogs to live in. The drainage canals with bare soils provide that habitat. Once the canals have concrete walls, water flows very quickly and a loss of habitat is the result (Lane and Fujioka 1998).

4.4.4 Satoyama Landscape Birds of Prey That Feed on Frogs

A number of birds of prey have satoyama landscapes as their habitat. Among these, Gray-faced Buzzards (*Butastur indicus*), which feed on reptiles and insects, are regarded to be at the top of the food chain in satoyama landscapes. Therefore, these birds are regarded as animals that indicate healthy ecosystems.

Buzzards are birds of prey that come to Japan (except Hokkaido) during summer to breed. In the areas of Inba Marsh and Tega Marsh, the buzzards usually come during early April and stay until October. These buzzards breed in south Siberia or north-east China and the northern part of the Korean Peninsula, from whence they

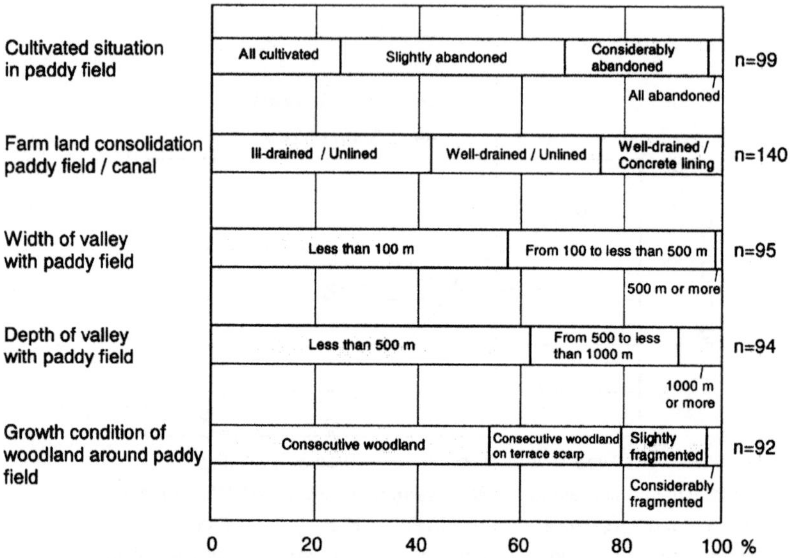

Fig. 4.15. State of paddy fields in breeding habitats of the Gray-faced Buzzard (*Butastur indicus*) based on a questionnaire survey (Azuma et al., unpublished)

arrive in early April staying through until the end of October before flying south. These birds winter near Okinawa and Taiwan, South China, Myanmar, Indonesia, Philipines and New Guinea.

Why do the buzzards fly to Japan? The reason is not known, but maybe the secret is in the satoyamas. A questionnaire survey on birdwatchers across Japan provided information on the life of the buzzard.

In Japan, it is known that the buzzards live in mountains up to 1000 m high. In general, most live in habitats of up to 100 or 200 m hills. In areas where it has been reported they are living, 80% of the places contain woodlands and paddies; 80% of these were yatsudas.

Figure 4.15 describes how the buzzard prefers yatsudas where, even though there are some abandoned paddies, you can also find cultivated paddies. The buzzards prefer places without drainage and with bare soil canals. Ninety-five percent of the valleys are narrower than 500 m and approximately 60% of the valleys are narrower than 100 m. Surrounding the yatsudas, you can often find woodland covering the slopes. If the woodlands are fragmented, the buzzards will not use them as a habitat.

Buzzards nest in trees, particularly pines. Ninety percent of buzzards nest on coniferous trees. They prefer trees that face the paddy fields. Nesting sites were identified and all were places that faced the yatsuda. The group surveyed 22 yatsudas in study area and identified a relationship between the existence of buzzards and land use. The study group concluded that the yatsudas where the buzzards live have woodland coverage that is significantly larger than the yatsudas without buzzards. The area of paddies that had been abandoned compared unfavorably in these terms (Azuma et al. 1999)

Therefore, buzzards prefer an area where conifers dominate the woodland. Also, the the valley has to be very narrow. Quite often, narrow have not had land improvement, meaning that, quite often, the valleys have drainage canals with bare soil.

4.4.5 Gray-faced Buzzards in Satoyama Landscapes

Now we know that buzzards are using the habitats, including the yatsuda, but how? Research groups have used radiotransmitters on male buzzards and have followed how the males have used the yatsuda during the breeding season (Azuma et al. 1998). The breeding season is from April to July. April is for making nests and, in July, the chicks leave the nest. Incubation is done by females and feeding is done by the males. Males made nests on *C. japonica* on terrace scarps. Feeding was a

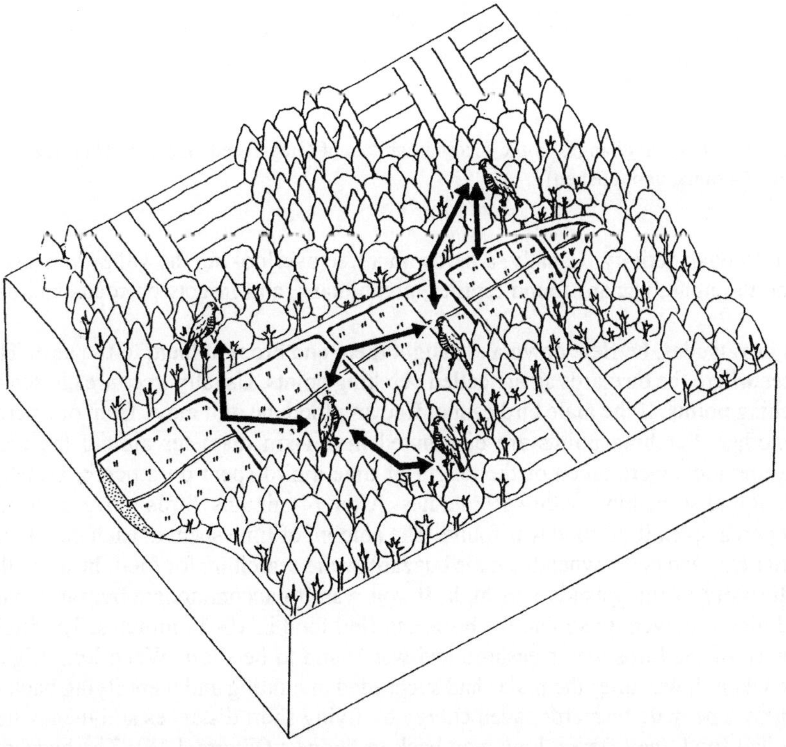

Fig. 4.16. Feeding behavior of the Gray-faced Buzzard (*Butastur indicus*) in the satoyama landscape (After Azuma et al. 1998)

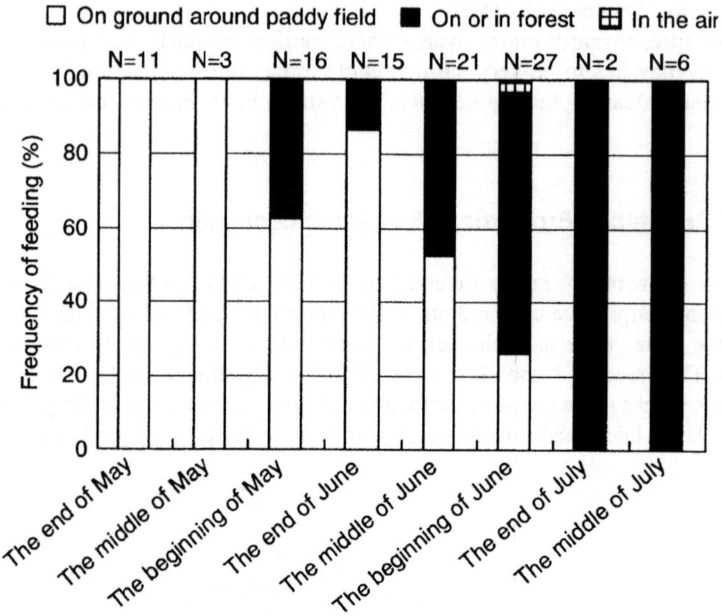

Fig. 4.17. Seasonal changes in the feeding habitats of Gray-faced Buzzards (*Butastur indicus*). (Azuma, unpublished)

dawn-to-dusk activity. Usually, a perch on a tree overlooking the valley was used. When the male buzzard found reptiles, amphibians, and insects on roads and pad

dies in yatsudas, he flew down and caught them and brought them to the nest. The places where the buzzards sit are called perching points. Usually trees are chosen as perching points. If the male bird cannot find any food, he will move from one perch to another. Perching points are distributed in a 1-km diameter around the nest. Measurements were taken of the length of time a bird spent on a perch. Usually, birds found something within 10 minutes. After 14 minutes, birds move on to the next perch, even if no food was found. The amount of time spent at each perch was similar for each perch when the male buzzards were searching for food. In short, the biodiversity of the yatsuda was high. If you want to accommodate buzzards, you need high biodiversity so that the birds can find food in 10–14 minutes. The flight distance of the birds was measured and was found to be short. When long flights were taken, it was after the males had succeeded in hunting and were flying back to the nest. The male buzzards saved energy by flying short distances and then, when they had food, they flew a long way back to the nest (Figure 4.16). The buzzards changed feeding places according to the season (Figure 4.17). In mid- to early May, after planting, the buzzards fed inside the paddy. After late May, when the rice is taller, it became difficult to hunt and the buzzards started to hunt in the woods. In early July, 100% of the time was spent in the woodland. Therefore, immediately after planting the rice, it became possible to hunt. When the rice started to grow, it became impossible. So, the buzzards moved onto the woodland part. The amount of small animals eaten was counted. The number of frogs that was fed upon decreased from May to July, whereas the consumption of insects increased during this time.

The height of the grass where the buzzards hunted was measured and found to be significantly shorter than in other places. So, the buzzards seek easily visible prey.

In concluding how the buzzards is adapted to the structure and function of satoyama landscapes, it has been found that the buzzards nest on trees overlooking the valley and perch mostly on trees with good sight lines. First, they feed on frogs, which can be found in paddies, and, when the grass becomes taller, they feed on insects on the leaves in the woodlands.

The yatsudas are very important during the spring, when food in other places is scarce. The buzzards prefer paddies that are cultivated because uncultivated paddies have high weeds.

Satoyama landscapes have all that is required for buzzards to eat, live and breed, and can be considered an ideal habitat for them.

4.4.6 Current State of Gray-faced Buzzards and Their Conservation

According to a questionnaire, 90% of the habitats for the Gray-faced Buzzards were found to be privately owned and 75% have no development protection. This means that the places in which the buzzards are now living are under threat of development. According to one survey, 60% of areas contain decreasing numbers of buzzards. This is due to the deforestation in south-east Asia, where the buzzards winter. However, in addition, development in satoyama landscapes may be a significant negative factor for buzzard numbers.

Changes in the satoyama landscape may bring about a significant impact on the wildlife of buzzards and frogs, which depend on a landscape that has been created by humans.

5 Approaches to Satoyama Conservation

5.1 Nationwide Partnerships for Satoyama Conservation

S. Nakagawa

5.1.1 Citizen Movements in Nature Conservation

The traditional use of coppices that have maintained woodlands over several hundreds of years has been destroyed by the fuel revolution of the 1950s and 1960s. During this time, fuel has changed from wood to fossil fuels. This period coincides with the beginning of the economic boom in Japan.

Table 5.1 Classification of woodlands according to age and the area of various types of forests and woodlands around satoyama and mountain zones in Kanagawa Prefecture (1995)

Elevation	Forest zone		Broad-leaved forest	*Cryptomeria japonica* plantation	*Chamaecyparis obtusa* plantation	Total area
300–800 m	Zones of supply forests and woodlands	Total area (ha)	18 722	11 670	7796	40 054
	(Higher mountains)	Ratio (%)	46.7	29.1	19.5	95.3
		31–35-year-old woodland (ha)	2415	2663	966	6058
		Ratio (%)	39.9	44.0	15.9	99.8
Less than 300 m	Zones of coexistence with human settlement	Total area (ha)	20 813	5963	1757	31 244
	(Satoyama landscapes)	Ratio (%)	66.6	19.1	5.6	91.3
		31–35-year-old woodland (ha)	4025	1146	216	5476
		Ratio (%)	73.5	20.9	4.0	98.4

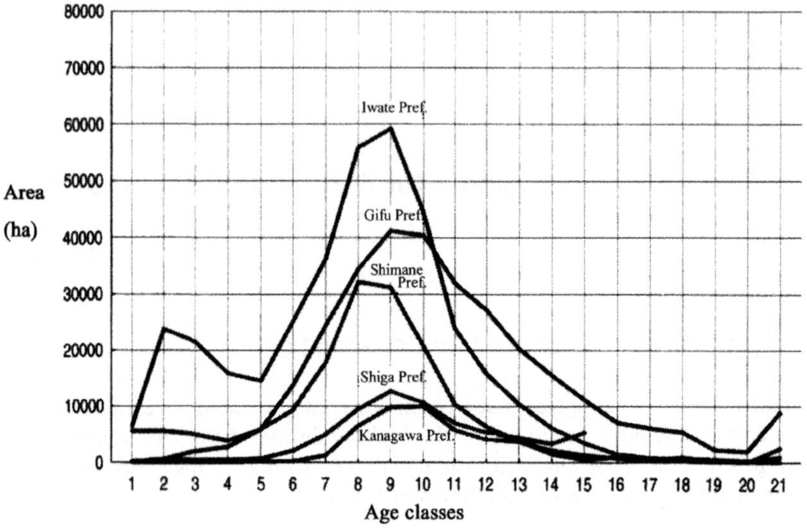

Fig. 5.1. Distribution of area of age classes of non-coniferous woodlands in five prefectures

Another major change with regard to the woodlands of Japan was the expansion of plantations of *Cryptomeria japonica* and *Chamaecyparis obtusa*. These species have been widely planted since the 1960s as projects for increasing the production of building materials were undertaken by the Japanese Forestry Agency. As a result, Japan managed to plant 1 000 000 ha in a very short period of time. The introduction and spread of the use of fossil fuels and chemical fertilizers removed the reason for the existence of the satoyama landscape, which was as a natural supplier of fuel and fertilizer.

Table 5.1 illustrates natural resource composition at different elevations in the landscape. In Kanagawa Prefecture, woodlands in areas up to elevations of 300 m have been designated as zones of coexistence with human settlement. These are typical satoyama landscapes. Woodlands in areas at elevations between 300 and 800 m have been designated as zones of supply woodlands, as well as *C. japonica* and *C.obtusa* plantation areas. They have been designated as woodlands whose main focus is resource production. Woodlands in areas at elevations over 800 m have been designated as zones of ecosystem conservation.

Within the satoyama landscape areas, there are 20 813 ha in deciduous woodland, and within coniferous plantations there are 5963 ha of *C. japonica* and 1757 ha of *C. obtusa*. The percentage of the total woodland area of 31 244 ha comprising deciduous forest, *C. japonica* plantations, and plantations of *C. obtusa* is 66.6%, 19.1%, and 5.6%, respectively. The proportion does not differ much when

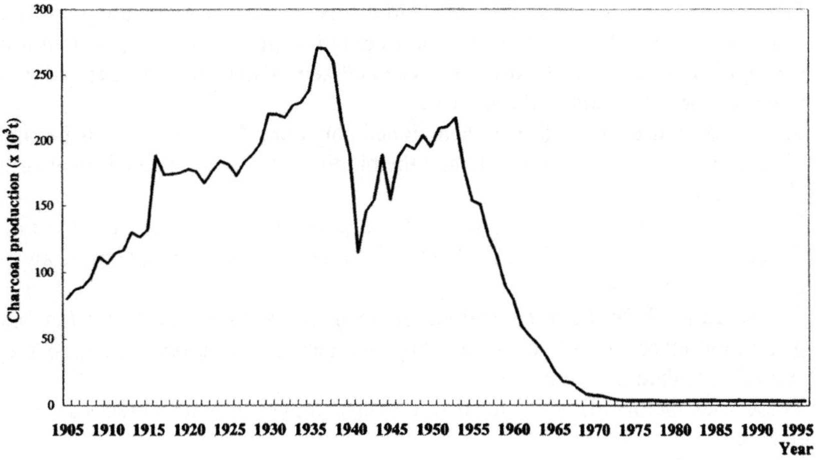

Fig. 5.2. Annual production of charcoal in Japan (Inamura and Nakagawa 2001)

taking into consideration only the 31–35-year-old woodland. The respective percentages for these areas in this woodland are 73.5%, 20.9%, and 4.0%.

On the other hand, in the zone of resource-supply woodlands, there were 18 772 ha of coniferous woodlands, 11 670 ha of *C. japonica* woodlands, and 7796 ha of *C. obtusa* woodlands within the 40 054 ha of total woodland area (46.7%, 29.1%, and 19.5%, respectively). The comparative proportions in the 31–35-year-old woodlands were 39.9%, 44.1%, and 15.9%. It is apparent that broad-leaved forests and woodlands occupy less area, whereas *C. obtusa* woodlands occupy more area, in resource-supply woodlands. It is also apparent that conifers were planted more in the zone of resource-supply woodlands; in other words, in the higher mountain areas.

Thus, satoyama landscapes were not subject to plantations and were essentially abandoned.

Figure 5.1 illustrates the transition of selected coniferous woodlands within different age groups. The woodlands have been classified into 5-year age categories (age-class) for ease of analysis. The area of extraction had its peak in the 1950s and 1960s, and decreased sharply after that. This implies a sharp decrease of wood resources from satoyama landscapes. This decrease is more significant in highly urban areas and less so in traditional charcoal-producing areas.

The neglect of the satoyama landscape can also be explained through the transition of charcoal production in Japan (Figure 5.2). Annual charcoal production, which peaked at 2 699 000 tons in 1940, decreased to 28 000 tons in 2000, a nearly 99% decrease.

There are several reasons why coppice woodlands have become popular since the 1980s:

1. During the 1980s, not only satoyama landscapes, but also many nature protection movements, came to the fore in Japan, triggered by plans to develop wetlands in the highlands. This was the occasion when citizens also realized that they had to work for the protection of these areas.
2. Satoyama landscapes had been abandoned for more than 30 years. It became obvious that the plant community and the biodiversity had changed dramatically since the war.
3. The growth in awareness of global environmental problems, such as acid rain, the greenhouse effect, and the depletion of ozone layer, made people more aware of environmental issues.
4. The so-called baby boomers started living in the suburbs. This was the first generation since the war to have a high interest in nature protection, so they started with their surroundings.

These four factors all contributed to citizens' movements to protect satoyama landscapes.

5.1.2 Grass-Roots Forestry Support Programs

When one talks about public involvement in the maintenance of the woodlands in the late 1980s, there are two major trends: one is the supporting program in the mountains; and the other is the conservation of satoyama landscapes in or around urban areas.

First, the public involvement supporting forestry. In 1985, a program called Woodland Maintenance Crusaders was started in Toyama Prefecture. This involved volunteers going to the woodland in Toyama and cutting the weeds. This innovative program consisted of a pioneer movement of mostly student volunteers. Following this, programs started in Tokyo, Shimane, Kagawa, and Kochi, and spread all over Japan and remain in operation today.

During the 1980s, another interesting project started in Miyagi Prefecture. This was initiated by oyster farmers. The issue was that the oysters may be farmed in the sea because of nutrients that are brought from the woods upstream. The oyster farmers organized a group to protect the deciduous broad-leaved woodlands. Again, a similar story followed in other places, such as Hiroshima, an area that is also famous for oysters. A third general aim of these programs is related to drinking water. This started in Yamanashi Prefecture and Kanagawa Prefecture because the drinking water in Yokohama is taken from the Katsura and Sagami Rivers, which flow through these two areas. So, a campaign to preserve the woodland was started to safeguard Yokohama's drinking water. A similar story followed in Osaka and many other places in Japan.

5.1.3 The Birth of the Movement to Save Satoyama Landscapes

The movement to save satoyama landscapes started in the 1980s, but can be divided into two major groups. One is mostly concerned with the maintenance of the coppice, the other is concerned with the maintenance of the plantation woodland, agricultural land, and irrigation ponds.

The ultimate goal of these two groups is almost the same, but the process is different. In terms of the group with a coppice interest, their focus is on the maintenance of the woodland by thinning, raking the woodland bed, and raising awareness about the maintenance of the woodland (e.g., through lectures). Education is the major purpose. In Kanagawa Prefecture during the 1980s, there were 23 groups working on 23 ha in total. These groups are mostly led by local municipalities. In Kanagawa, there are also a number of NPOs that have been working on the maintenance of and education about these coppice areas. In terms of those types of groups that include other elements, there are two outstanding NPOs: Friends of Maioka Park in Yokohama, and Friends of Shishitsuka.

In Maioka, the citizen volunteer groups conserved the area. Yokohama City today collaborates and is partly contracting the maintenance of the park through the volunteer groups. Yokohama City asked the volunteer groups first to conduct the maintenance of the paddy in the park, recreational cultivation and the maintenance of the satoyama landscape itself. In terms of the Shishitsuka group, they have been the leader for groups in all Japan. They organized a conference called the Satoyama Summit in 1982, and collaborated with ecologists and local historians. The Shishitsuka group not only works to maintain the satoyama landscape, but also gathers a great deal of information on how local people are maintaining the area.

5.1.4 Local and National Policy

Even though a large number of groups have been at work, it is sometimes difficult for these groups to push the maintenance forward and they sometimes require support. There is a problem of gaining rights, as well as access, to land. In many cases, accidents and trouble have been reported, such as conflicts with landowners and accidents with chainsaws. Because of these difficulties, local municipalities and the government realized, in the 1990s, that they should support the NPOs. For example, in 1992, the Forestry Agency started a program called Creating a Woodland With Citizens. Many local municipalities copied this program. In these programs, not only was the maintenance of the trees emphasized, but so were maintenance activities, such as raking, and recreational activities. Some of the pioneering programs of local municipalities for the conservation of satoyama landscapes include the program of Yokohama City, in 1971, Woodland for the Citizens, and another in Sagamihara City, Kanagawa Prefecture, in 1973. These programs have had a tremendous effect on the education of leaders (e.g., City has had 24 groups led by alumni of the program on 452 ha of woodland).

Table 5.2 Conservation activities of woodlands at different stages

	Time period	Activities	Viewpoint changes and social relationships
Stage I	1–3 years	Weeding of abandoned woodland floor etc.	Temporal and bulk biomass processing, woodland recreation
Stage II	3–5 years	Reconfirmation of effects of woodland floor management	Effective vegetation management
Stage III	5–10 years	Stable woodland management	Enlargement of management
Stage IV	More than 10 years	Stable woodland use	Socialization, use of biomass and maintained woodland space

5.1.5 Collaboration Among Scientists, Citizens, and Planners

It is important to realize that the involvement of citizens should be supported by science in forestry and the environment, as well as by specialists in planning, artists etc. To enable this collaboration, there should be several occasions on which representatives from all sectors sit around the same table. This can be achieved by having seminars, or to have specialists as consultants for the activities from the beginning. In many cases, the activities conducted by the citizens have gone wrong because they have received the wrong information through the mass media. Mistakes can also happen where there is a lack of good communication between stakeholders. In these cases, the professional expertise of scientists or planners is very helpful, which it is why it is necessary.

5.1.6 Improving and Documenting Activities

As described in Table 5.2, even in one single NPO the purpose of the activities may change over time, which is understandable. So, this means that it is necessary to record the group's activity over time. In addition, it is necessary to improve the group's activity over time by having a record. However, the ultimate goal of the activities of all these groups is to have a well-maintained satoyama landscape as a natural or cultural heritage of the area. In historical times, such areas did have significance for the public, because the land area was regarded as common and had the function of conserving the surrounding landscape of paddies and crop fields. Anyone wanting to harvest firewood, mushrooms, or *Pleioblastus chino*, was able to do this in the satoyama landscape. In some cases, there were some restrictions (e.g., in some satoyama you could not harvest the highly prized *matsutake* mushroom). The NPO groups should also provide and store information for the general public to encourage a better understanding of such areas. It is also necessary to receive information from and acknowledge the experience of other groups. For these reasons, many of the groups currently have their own websites, as well as newsletters.

In recording the history of the group's activities, it is necessary to have a maintenance map or a manual that can be used not only by the group, but also by the general public.

5.1.6.1 Maintenance Map

To have adequate maintenance of trees in satoyama landscapes, you should have a record of each tree. Such a maintenance map will also be very useful in protecting endangered species (Figure 5.3). A number of NPOs have made their original maintenance manuals but, in some cases, local municipalities help to make these. A good example of a manual for the coppice woodland is that written by Kanagawa and Aichi Prefectures (Figure 5.4).

5.1.7 Partnerships Between NPOs in Japan

Most NPOs working on the conservation of satoyama landscapes were established in the 1980s and 1990s. Today, approximately 500 groups exist. Although some linkages exist between the groups (such as the exchange of information), it can be said that it is necessary to have an umbrella organization for all these disparate groups. As a way of achieving this, in 1992, the National Coppice Woodland Congress was established and, in 1993, a nationwide conference on the citizens and woodland was organized.

There have been nine annual conferences in different parts of Japan. The major topics that have been discussed at these conferences have been maintenance methods for coppice woodlands, how to make links with local municipalities, and how to have successful activities. More recently, the Congress has had a keen interest in therapy in the woodlands. In addition, the Congress has published several books on coppice woodlands in Japan.

A congress on Citizens and Woodlands has taken place every year since 1993, with the main topics being forestry, problems related to the ageing of society, and overseas volunteer activities.

The third nationwide forum is the Yokohama Woodland Forum. This was established in 1997 and consists of 30 NPOs found in Yokohama City. This umbrella forum has been working on organizing seminars and lectures, and on supporting leaders.

Forest 21 was established in 1994 in Kanagawa Prefecture and has worked on the training of citizens in maintaining national forests and woodlands. This organization conducts training courses in the maintenance of forests and woodlands in schools or private companies.

Apart from the umbrella groups, much has been achieved through partnerships. Among these groups, some have been incorporated, are now cooperative organizations, and now lead the umbrella organizations. For example, in Shimane Pre-

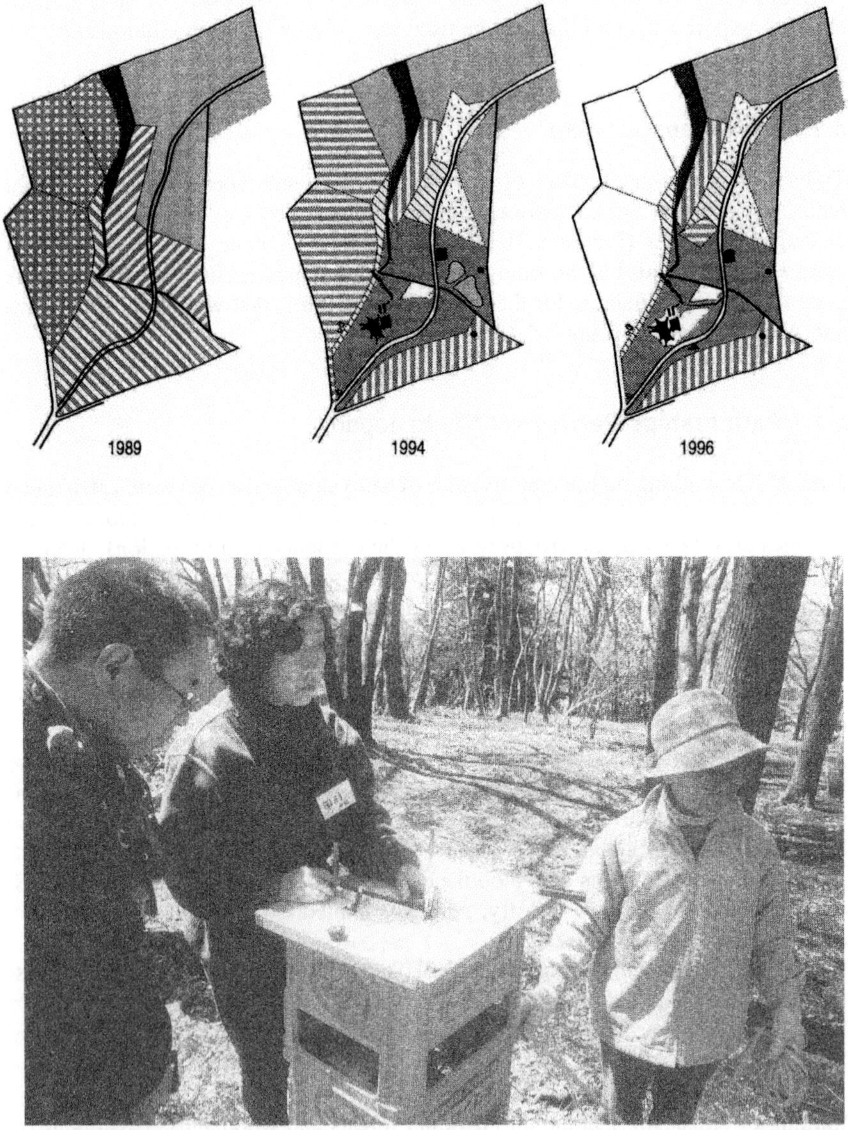

Fig. 5.3. Annual changes in vegetation (upper) and mapping workshop for the management of vegetation in the woodland (lower)

Fig. 5.4. Manuals for woodland management

fecture, a group called Hiigawa Club, a cooperative, is leading many other satoyama landscape conservation groups in the prefectures, as well as organizing many symposia and conferences. The Hiigawa Club is also involved in promoting local industry in the prefecture. Such an umbrella organization may become very important in developing a better relationship between the groups and the public.

5.2 Coppice Woodland Maintenance by Volunteers

N. Kuramoto and Y. Asou

5.2.1 Coppice Woodland Volunteers at Sukuragoaka Park

Sakuragoaka Park is located on the eastern edge of Tama New Town, Tokyo. The volunteers work on 2 ha of coppice woodland known as *Konara-no-Oka* (Oak Hills). The volunteer activities started in 1991 and, every year, approximately 80 volunteers have registered. Every second Saturday the volunteers work on the satoyama landscape. Their work includes maintenance, survey, lectures, and training (Kuramoto 1994, 1996. Figure 5.5). As well as maintaining the woodland, the

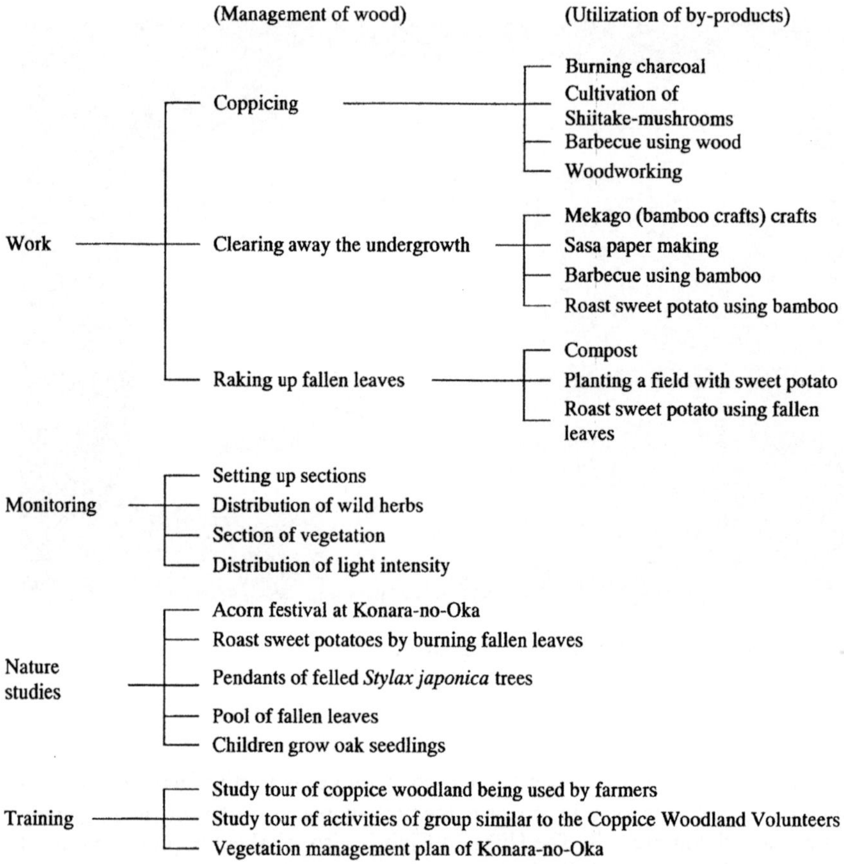

Fig. 5.5. Programs of the Coppice Woodland Volunteers at Sakuragaoka Park

volunteers utilize the organic matter. Although the area is only 2 ha in size, more than 200 people have worked on these woodlands, reflecting the attraction of the place for local citizens. When the volunteer program was started in 1991, most of the citizens had no knowledge or prior experience of maintaining coppice woodlands, but applied as volunteers because they wanted the exercise. Gradually, as they worked on the woodlands, they started to realize the importance of what they were doing, and started to find the work more interesting. Their activities are based on the *do first, think next* concept. Let me explain how this works.

Because most volunteers who first came into the program had no idea about the organization and maintenance of coppice woodlands, in order to organize and generate a common understanding, you first had to let them try out a technique.

Through a pilot trial, the volunteers gradually come understand the nature of the coppice woodlands, and the importance of the maintenance of these areas. The first maintenance work volunteers start with is the cutting of *Azuma nezasa* (*Pleioblastus chino*). When these plants are being cut, it is necessary to imagine what kind of woodland bed is needed as a target. The volunteers need to have clear idea of what landscape they are working towards. That type of image helps people to work toward a common purpose. When this cooperative work starts, volunteers can move onto the cutting of bushes and low trees. At this point, volunteers can also work on raking the woodland bed. When all the members of the group have their own idea of the woodland bed they want to see, the next stage is to share these ideas among the group. To do this, it is necessary to encourage each member to talk and share information and ideas. When the ideas have been shared, the volunteers start to alter the coppice woodlands toward the target vegetation, and then organize meetings to discuss maintenance plans to sustain the target vegetation. At this stage, local municipalities collaborate with the citizens so as to have the maintenance plans conform with other plans of the municipality.

It is important when you are starting the whole procedure that you focus exclusively on woodland bed maintenance: the cutting of weeds and *P. chino*. If you make a mistake, then you can try again. Cutting down trees can, potentially, inflict more damage.

5.2.2 Monitoring Citizens

Even when volunteers have a common goal and plan, it does not mean that nature will recover as they thought. In many cases, there will be a gap between the goal and reality. The plan has to be revised. To do this, monitoring is necessary. The criteria of the monitoring process should be pretty broad: when flowers are blooming, how hard the soils are etc.; this all depends on the goal. The monitoring process undertaken by citizens cannot be based on analysis with special skills or instruments. On the other hand, the monitoring process undertaken by citizens has advantages in the number of people working and the ability for monitoring to be conducted on a daily basis.

As a result of this monitoring process, it may be necessary to revise the maintenance plan. The revised goal should be shared and the responsibility of each citizen should be defined. This does not mean that special skills are called for, but what can be achieved independently in the long term, without a lot of stress, should be defined. Because, in many cases, the goal of maintenance will be achieved over a period of more than 10 years, maintenance works should be thought of on a very long-term basis. The citizens must understand that their responsibility will continue for more than 10 years. If the work cannot be enjoyed, the people will become disillusioned and work will stop.

When advertising for volunteers, you should not force them to register but wait for them to register. There are only two criteria for this: (a) the volunteers must show an interest in coppice woodlands; and (b) everything should be their own responsibility. Also, the work may become better if there is a range of ages and work that the volunteers can be involved in. The activity will become richer and more sustainable. Insofar as the volunteers fulfill the two criteria above, many different citizens should be encouraged to come and work with the group. Aged citizens and children may have specific work reserved for them so that all people can participate in the group's activities. The Coppice Woodland Volunteers at Sukuragoaka Park ranged in age from teenagers to old-age pensioners. Most were from Tama City, but there were also some people from East Tokyo. In terms of their occupation, you can find specialists among the volunteers, such as landscape consultants, but also university students, housewives, and older people.

5.2.3 History of Vegetation Maintenance by Citizens

The 10-year history of vegetation maintenance at Sakuragoaka Park can be divided into three eras. From 1991 to 1993, the group started to organize a common idea of the target vegetation of Konara-no-Oka. At this time, they first tried to share the idea through training courses and excursions. The group started to cut the trees for encouraging saplings to grow, and for stimulating regeneration of woodlands. From 1994 to 1997, there was generally extensive maintenance after first thinning the woodland . During the period 1998–2000, the group realized that if they thinned the woodland there would be a too-drastic change in the vegetation, so they refocused the whole procedure based on the reality they observed, using the *do, think, observe, and do again* principle.

5.2.3.1 Details of the Three Eras

In 1991, the group started with maintenance, despite mistakes, of mostly raking the woodland bed. Also, excursions and courses were held to set a maintenance goal. In 1992, the group set the goal of raising the biodiversity of the coppice woodlands, as

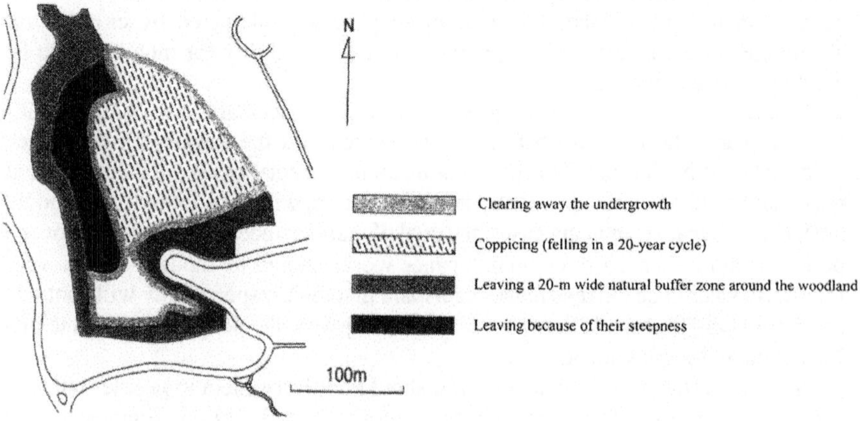

Fig. 5.6. Vegetation management plan of Konara-no-Oka

well as a goal of maintaining the coppice landscapes that used to be there when the plant matter was used by farmers. The third goal was to start environmental education. The first goal was set for deciduous-leaved young woodland, which restricts the growth of bushes. To facilitate this, the group divided the whole area into nine blocks. Then they thinned each block, every couple of years, in rotation. For each block, after felling the trees, the group maintained the woodland by cutting weeds and *P. chino* every 1–2 years (Figure 5.6).

5.2.3.2 Felling Trees

In 1992, the group first felled the back part of an area of 200 m^2, located at the end of the experimental site. When felling large trees, the group needed a specialist. In 1993, the area cleared was planned to be approximately 400 m^2. In 1994, the group cut 800 m^2 and, in 1995, they cut 2200 m^2 of the planned area or 3000 m^2 in total in 1994/95, the original size proposed for felling over 10 years (Figure 5.7).

This occurred because each subgroup worked on the trees in isolation and did not communicate very well with the other subgroups (there were lots of small groups). Finally, in 1999 and 2000, a moratorium was called on the felling of trees.

5.2.3.3 Woodland Bed Management

In the vegetation maintenance plan of 1992, it was planned to have woodland bed management every year. In 1994, the group realized that this was not necessary and that the bed could be left for 2 years. In the end, the woodland bed was managed this way. Also, the group became busy attending conferences and other events, and so were not able to maintain woodland bed so often. More than 70% of the area was abandoned for more than 5 years. The bed management of the area that had been clear cut in 1992 started again in 1996 and did not last very long. The *P. chino* recovered so quickly that, within 3 years, it had recovered to a greater extent than what it used to be. The area that was cut in 1995 suffered the most extreme growth of *P. chino* because, in only 3 years, the average height of the *P. chino* was as much as 2 m and, in some cases, reached 6 m (Figure 5.8). Therefore, the *Q. serrata* sprouts died and none of the flowers of the grasses bloomed. At that stage, the volunteers realized the seriousness of the situation and maintained the woodland bed. Before 1996, the total area maintained was less than 600 m^2; then, between 1998 and 2000, the managed woodland bed was clear cut. The volunteers stared to manage this after 4–5 years of abandonment. On the lower part of the hills, the volunteers could identify some species of flowers and some species of sprouts growing. On the upper part of the hills, the situation was not so clear (Figure 5.9). In areas where the recovery of the grasses, flowers, or sprouts could not be identified, the natural regenerative potential of the area had been sacrificed as a result of previous malpractice. This kind of mistake may be fatal for an area such as the hills, which are isolated from their surroundings, producing an island effect.

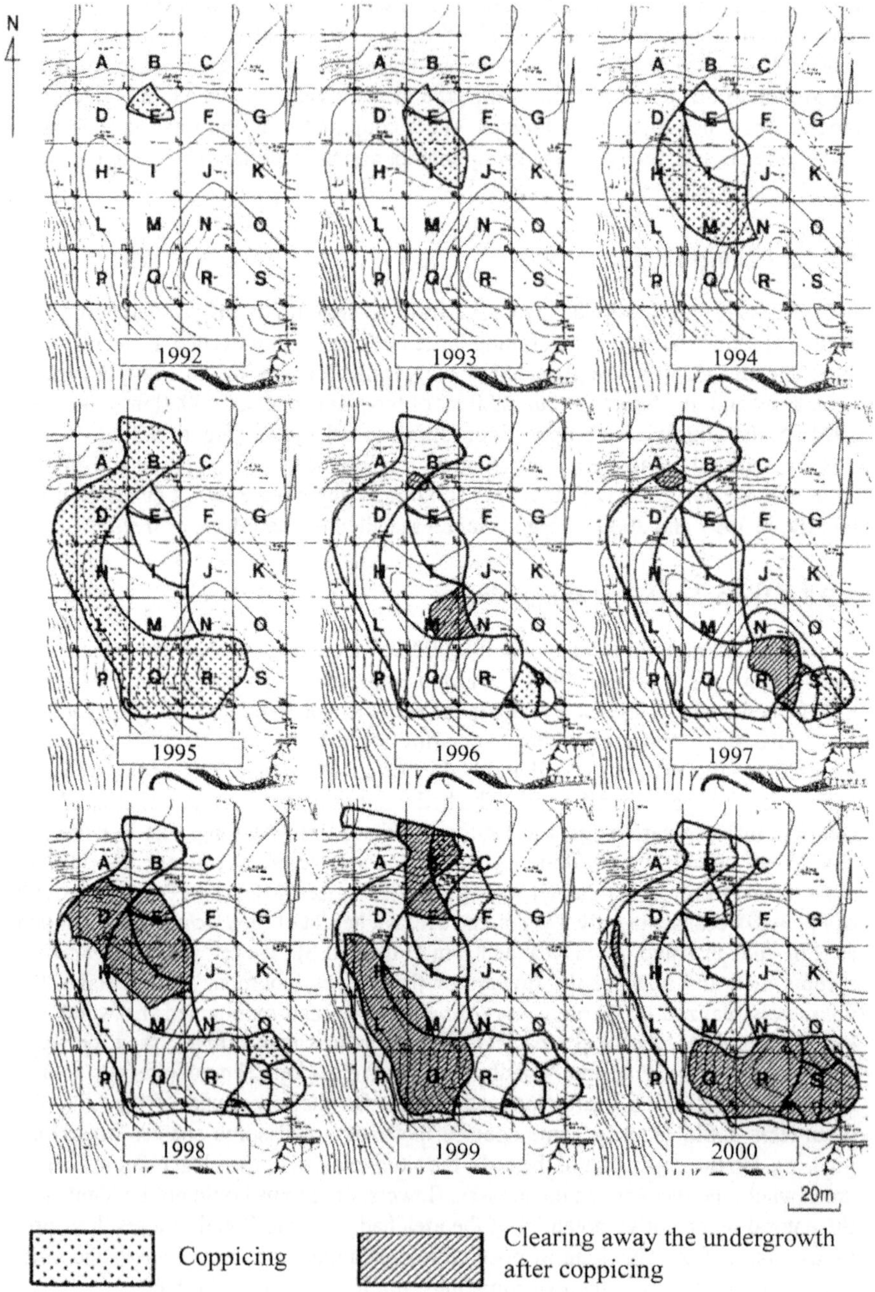

Fig. 5.7. Progress of coppicing work in Konara-no-Oka

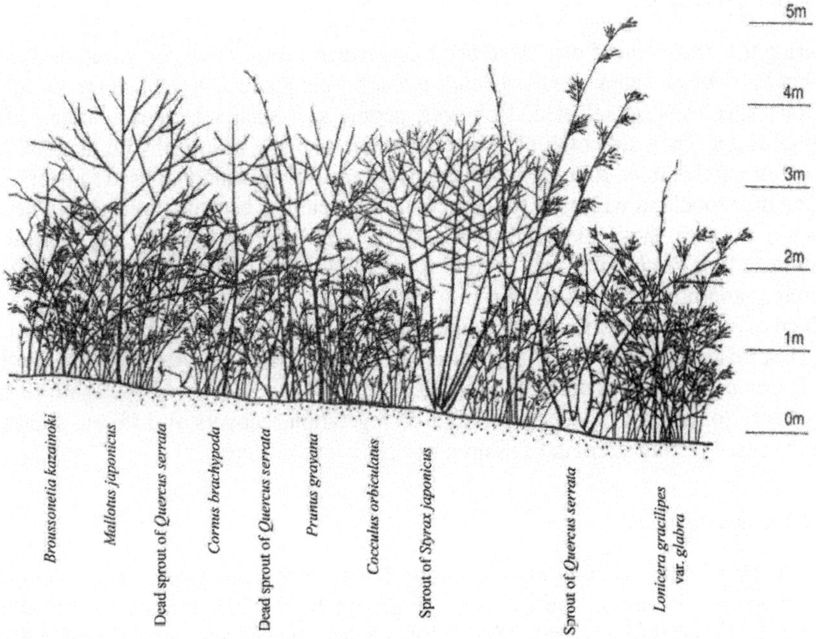

Fig. 5.8. A cross-sectional representation of vegetation 3 years after cutting. The *Quercus* sprouts cannot compete successfully for sunlight

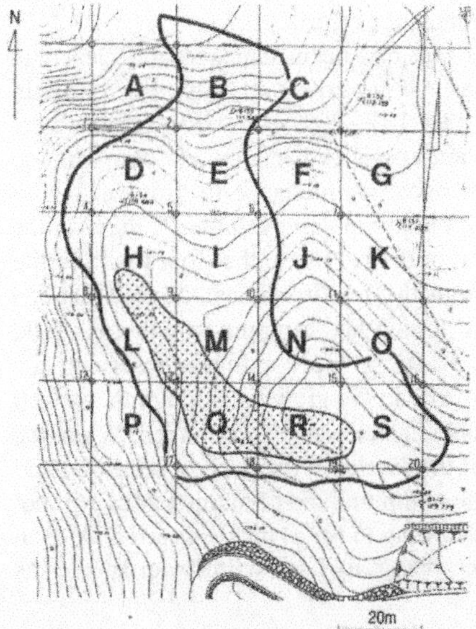

Fig. 5.9. The coppice volunteers cleared away the undergrowth repeatedly in the *dotted area*, but the wild herbs have not recovered

5.2.3.4 Thinning

During the first cycle of woodland bed management, from 1996, the group decided to cut *P. chino* and some berries (*Rubus palmatus* etc.), and also to plant trees. Only certain species were selected. However, people saw such selection as being too complicated. They did not think of the landscape as being beautiful with that. What they ultimately aimed for was to restore the *Quercus* woodland that used to be there when the woodland was used for charcoal. Clear cutting became increasingly necessary, because those places that underwent clear cutting became aggressively dominated by *P. chino* again. Therefore, in 2000, the group realized the need for thinning and not clear cutting, but only removed species such as *Pueraria lobata*, which may come to dominate the whole area and restrict the growth of *Q. serrata*.

Despite all the precautions taken to ensure that such thinning restricts the growth of *P. chino*, those plant species that prefer sunlight may not grow back again. It is necessary to monitor closely what kind of vegetation recovers at different stages, and to fine-tune the strength of thinning or cutting accordingly.

5.2.3.5 Monitoring

For the first 3 years, monitoring was conducted by encouraging volunteers to enjoy the coppice woodlands. Then, during the period 1994–1996, monitoring stopped because of a shortage of volunteers. In 1997, the area had become so overgrown that it became to difficult to monitor and identify which part had been cut and when. To make monitoring easier, a map was made in 1997 that divides the whole area into 20 m grid cells, with monitoring performed within the grids. The aim of this was to record what had been done in each section and to feed this into the next work plan for the area. To achieve this, it was necessary to record the history of thinning or clear cutting of woodland bed management. It was also necessary to record the amount of work according to different maintenance efforts.

In 1999, the group recorded the places that had been managed and clear cut. Through this exercise, they realized that half the *Q. serrata* sprouts were dying. In all areas except for the tops of the ridges, only 10%–20% *Q. serrata* sprouts were still alive. In general, the growth of the sprouts is rather slow. If you want to maintain an *Quercus* sprout, you need intensive maintenance to provide it with enough sunlight. Also in same year (1999), the group surveyed the *P. chino* in the area and identified that the slopes to the west or the bottom of the slopes with fertile soil had *P. chino* with good growth. The group also confirmed that the area that had been abandoned for a long time after clear cutting had returned to a jungle. Wild flowers were very few in number. In 2000, the group continued observing the growth of the sprouts of the old trees and recorded the growth each month, as well as changes in the vegetation in areas that had been clear cut and thinned. When the height of the saplings exceeds the height of the *P. chino*, the purpose of woodland management is to conserve wildflowers. The optimum kind of maintenance will not be the same for all species, but will vary according to the plant species. To have an optimum maintenance procedure or a target, the group used wildflower species as indicators;

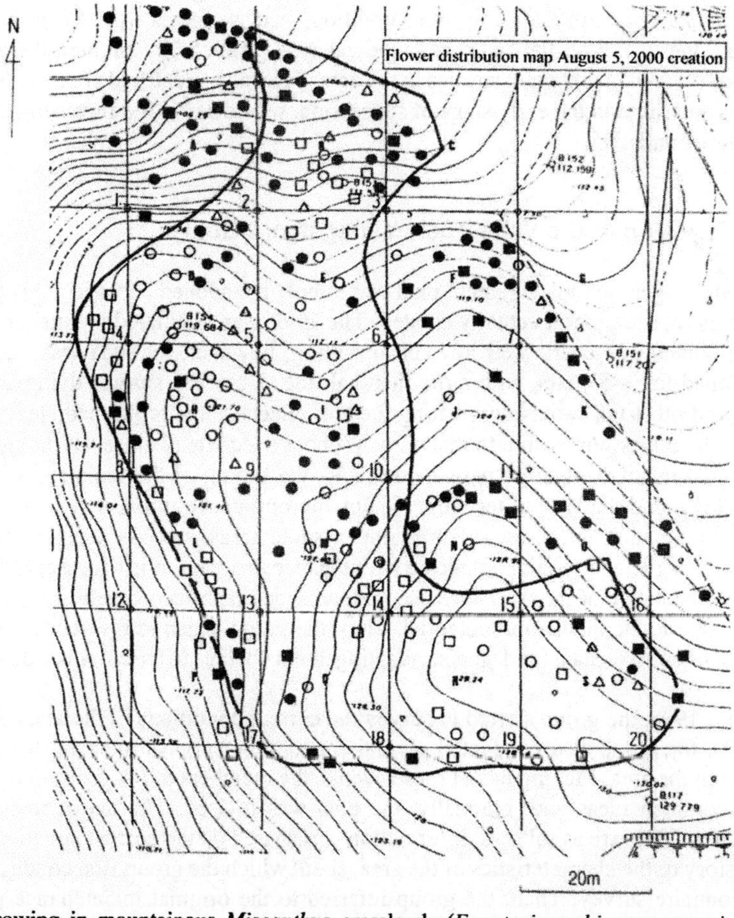

○, Growing in mountainous *Miscanthus* grasslands (*Eupatorium chinense* var. *simplicifolium*, *Lysimachia clethroides* etc.)
□, Growing in managed coppice woodland (*Carex nanella* etc.)
●, Growing in deciduous woodland (*Tricyrtis macropoda*, *Chloranthus serratus* etc.)
■, Growing in evergreen forest (*Cymbidium goeringii*, *Liriope platyphylla* etc.)
△, Others (herbs on woodland edge and landslide in mountain district)
The *open symbols* designate herbs that like sunny spaces; *closed symbols* designate species that do not like disturbance from excessive management or strong sunlight and high soil temperatures in the summer.
This map shows individuals and populations that grew to prominent sizes.
Most of J and K were not examined.

Fig. 5.10. The coppice volunteers cleared away the undergrowth repeatedly in the *dotted area*, but the wild herbs have not recovered

for example, *Adenophora triphylla* indicated well-managed light conditions and *Tricyrtis macropoda* indicated stable conditions. Grass species were found mostly on the ridges and wildflowers were in the valleys (Figure 5.10). In total, the group counted more than 400 plant species in the area. They also identified areas with vine species and those in the early stage of succession, where herbs and flowering plants were few in number.

5.2.4 Revision of the Vegetation Management Goal

The 1992 vegetation management plan was simply abandoned because none of the members realized that it actually existed. The 1992 plan stipulated maintenance of the woodland bed every year after clear cutting; however, the area was, in fact, abandoned for 3–5 years, with more than half the *Q. serrata* sprouts dying and *P. chino* and other tall weeds dominating the area. Because the group had clear cut a wide area, it was impossible to maintain. Without adequate plans, even though the group had maintained some parts of the land, the timing of the maintenance, the technology, and the type of terrain were not appropriate, resulting in a serious impact on the nature in the area. For example, some areas, such as northern facing sites, are fragile and require a delicate maintenance program. It is necessary for the group to be aware of this and to revise maintenance plans accordingly so that they do not deteriorate the natural potential. Thus, the group began to revise the vegetation maintenance plans and goals, adapting them to the different situations that arose.

From 1998, the group started to survey the current conditions of the area. Then, in 2000, the group also conducted several cooperative surveys to share the information in the area. Making use of these events, the members of the group started to share common ideas and, gradually, the plan was revised. The major revisions, enacted in 2000, are as follows. Information was shared on the current situation and the history or the characteristics of the area, about which the group first conducted a questionnaire survey. Then, the group referred to the original maintenance plans and goals from 1992, and the subsequent changes. They then identified the problems of the 1992 plan and set a new direction for the maintenance of the area accordingly. This new direction was implemented in the maintenance plan.

In the future, based on this plan, it is expected that the natural potential of the area may be maintained through an appropriate cycle of work on the woodland, monitoring, checking, and planning. Having volunteers who applied to join the group through the newsletter and local advertisements meant that these volunteers were non-specialists and could not be aware of problems, even as they occurred. Another problem is that these volunteers tended to be more enthusiastic about cutting down trees and making charcoal rather than weeding. This could be seen in the maintenance plans, which emphasized the felling of trees and charcoal production rather than woodland bed management. It is necessary to educate people in order to fill the gap between what they have done and what needs to be done. If you clear cut abandoned coppice woodlands covered with a large amount of *P. chino*, the woodland bed receives a lot of sunshine. This causes *P. chino* and vine species to grow at an overwhelming rate. Once this happens, there is little to be done in terms of maintaining the area by hand. Once the area turns into a jungle, woodland bed management and surveys on sprouts become impossible. In addition, the growth of *P. chino* and vines will change the ecosystem of the woodlands. Grass species and flowering species will become extinct because their niche will be taken by vines and *P. chino*. This leads to a deterioration in the nature of the area and, so, the question has to be must be asked: what is the point of maintenance in the first place? After 10 years of vegetation maintenance in the hills, it has become clear how delicate an issue maintenance is, especially maintenance by volunteers. Maintenance has to be flexible enough to adapt to changes in the ecosystem while maintaining a good balance between the size of the area to be maintained and the amount of labor needed. Also, a very clear idea of the purpose of the maintenance and what is required to achieve this is needed. Without this, the system cannot adapt. Some of the maintenance activities are more attractive than others. However, volunteers have to realize what is necessary for the coppice based on their responsibilities and not on their preference for the work. To be successful, volunteers have to participate in the program at all levels and stages.

5.3 Regeneration of Satoyama Landscapes

Y. Kitagawa

The Zushi-Onoji Historic Environmental Conservation Area was designated as a Greenery Designated Conservation Area by the Tokyo Metropolis in 1978. The designated area spreads across the rolling Tama Upland to the north of Machida City, Tokyo. The area maintains biological integrity within an intricate mosaic of coppice woodlands and yatsuda areas, valley bottom rice paddies. Until May 2001, half the area was privately owned, while the other half was owned by municipal governments. The Tokyo Metropolis has entrusted the Machida Historic Environmental Management Organization with the vegetation management of a 16 ha area since 1996. The organization consists of local farmers and non-farmer volunteers.

It is no longer a new idea to have the management of coppice woodlands in municipal conservation areas carried out by regional farmers and residents. Regional farmers, who are very familiar with the land, are able to manage the coppices and rice paddies with long-practised procedures. There is no doubt that it is one of the best methods in terms of conserving satoyama biological diversity as secondary nature. This idea is possibly applicable countrywide. The following describes the circumstances for the establishment of the organization that is now maintaining the satoyama environments that are experiencing a revival after being neglected for a quarter of a century.

5.3.1 Historic Environmental Conservation and Citizen Movements

On 4 July, 1978, 33.2 ha of land that extends over Zushi and Onoji Towns in Machida City, Tokyo, was designated as the Zushi-Onoji Historic Environment Conservation Area under The Ordinance Regarding to Nature Conservation and Rehabilitation in Tokyo. Conservation of the area was first proposed in the early 1970s by interested citizens led by Mr. Hiroshi Saito, who endeavored for natural conservation of the Tama Hills. The proposal included the following conditions: (a) no trouble should be created for the local community; (b) there should be incentives in the form of tax reductions; (c) citizens should be well informed about the management; (d) an office should be provided for the person in charge; and (e) volunteer cooperation should be solicited. In December, 1974, the city proposed that Onoji Castle remain as a candidate, against the will of the local community. The new bylaw restricted activities such as the following:

1. New construction, reconstruction, or extension of buildings and other structures.
2. Residential development and land reclamation in any form.
3. Mining of minerals and quarry.
4. Landfill and reclamation by drainage.
5. Extraction of wood resources in the area.

Between September and December, 1977, explanatory meetings were held with landowners. These meetings, however, did not lead to a consensus. Led by Mr. Koichi Tagoku (currently Head of the Machida Historic Environment Conservation Management Organization), local residents submitted an opposing petition to the governor at the time, Ryokichi Minobe, on 20 December, 1977. The opposing reasons were as follows:

1. Tax relief was not provided.
2. Extensions of agricultural cultivation and woodland development were not allowed.
3. Limits on tree extraction were too restrictive.
4. Rice production control by the national government may necessitate paddy reclamation by drainage or field reformation.
5. No activity was allowed within 6 m of an agricultural road.

At the very end, the opposing petition submitted by local residents said:

> "... please do not sacrifice urban farmers for development. We have been in agriculture for generations. We dearly beg you to develop a designation so that we can work constructively with no worries.'

The confusion and resentment of local residents was evident on the basis of the contents of the petition. Mr. Tagoku later reminisced that he beetled to collect signatures because of his thought about local sustainability and his love of traditional agricultural landscape.

Several meetings regarding the conservation designation were held between residents and the municipality in order to consider the petition submission. According to an announcement from the Tokyo Metropolis, the conservation policy was determined as follows. The site of the Onoji Castle remains were to be preserved in their present condition. The surrounding woodland was to be managed in a manner so as to allow it to grow back to a woodland of native vegetation. Other woodlands were to be maintained in their current condition; that is, coppice woodlands composed mainly of *Quercus acutissima* and *Q. serrata*. For each yatoda, agricultural activity was promoted as before. For other neglected rice paddies, a diverse natural environment of wildlife habitat and ecology was intended. Existing topography was to be conserved while securing an inflow of water. Thus, wetland environments were created. Documents indicated a good understanding of the issues regarding the conservation and management of nature within the local area. Landowners were assured a right to claim access to land purchased by the municipality. However, as mentioned previously, this bylaw was intended to create restrictions under the national conservation policy and no incentives, such as tax reductions, were applied.

Local residents, who were requested to cooperate with the conservation, felt that they were forced to trouble themselves with one-way sacrifices. This was a time of a decrease in agricultural labor following the national rice production control and conversion of rice cultivation strategy from the 1970s. The area was within the urbanization control designation. The local farmers resented the fact that the area was being left behind while extensive developments, such as Tama New Town, stood out in the surroundings.

The then Department of Pollution (currently Bureau of Environment) of the Tokyo Metropolis continued to hold explanatory meetings with local residents and also with Mr. Tagoku. In the end, the bylaw was applied under the condition so as not to obstruct ordinary local life and to achieve the objectives of conserving nature and traditional land-use activities, including agriculture. Incentives, such as tax reductions, were applied in the Historic Environmental Area within the urbanization control area, based on Law Regarding to Public Land Expansion. Tax exemptions of US$120 000 (15 000 000 yen) were applied for special deductions on purchases.

On 25 May, 1978, the Nature Conservation Council Report was submitted, which designated the conservation area and responded to the concerns of the various parties. With the approval of the Mayor of Machida City, on 4 July, 1978, Zushi-Onoji Historic Environment Conservation Area was announced, 7 years after the first proposal was submitted.

5.3.2 Machida Historic Environmental Management Organization

Historic Environment Conservation Areas in Tokyo aim to conserve local historic buildings within their surrounding natural environments. Unlike public parks, which are intended for public use, conservation areas have restrictions on public access. The Zushi-Onoji Historic Environmental Conservation Area proposed to severely restrict public access, as well as agricultural activities. However, despite the initial intention, public intrusion into unfenced green conservation areas increased dramatically as a result of a large residential developments nearby. The theft of agricultural products and wild plants, the dumping of garbage, driving vehicles into cultivated fields, and other such issues of public morality frequently occurred. In addition, natural disasters (such as wind and flood damage) and artioccurred. In addition, natural disasters (such as wind and flood damage) and artificially caused disasters (such as woodland fires) have occurred. The anxiety of local residents has increased.

Correspondence, led by Mr. Koichi Tagoku, who was concerned about the issue, proposed, in September, 1992, the establishment of a management structure organized by local farmers in order to revitalize the local economy while conserving local natural environments. In addition, the Machida Historic Environmental Management Organization was established, aimed at vegetation management in the conservation area. Many local citizens who were interested in the issue, along with scholars and politicians from the local municipality, have supported the project.

Vegetation management of metropolitan property land is commonly undertaken by professional landscape contractors. It is difficult to have private contractors provide adequate management for each satoyama landscape depending on seasonal conditions. Satoyama landscapes, which consist of a land-use mosaic of rice paddies with water channels and footpaths inbetween, water reservoirs, fields, roads, and coppice woodlands based on intricate microtopography, are best managed following traditional procedures by local farmers.

Accepting this, Tokyo Metropolis conducted an investigation in 1995 to survey the existing conditions and to examine future strategies. The investigation made a comparison of fauna in two different yatoda environments, plus a description of traditional management procedures. One was an adequately managed environment and the other was an abandoned yatsuda environment.

Following the proposal and investigation, Tokyo Metropolis decided to entrust vegetation management of the conservation area to the Machida Historic Environment Management Organization in June, 1996. With the cooperation of the city, entrustment of the environmental management in the conservation area was made in October of the same year. Thus, the first attempt at environmental conservation by local farmers in Japan had started.

5.3.3 Regeneration of the Shinmei Yato

5.3.3.1 Restoration of the Yatoda

Management by the Machida Historic Environment Management Organization started in the autumn of 1996 in the Shinmei Yato with an area of 4000 m^2. The yatoda had been neglected for 25 years. Footpaths that had been structured through the paddy fields had collapsed and the area looked like a neglected *Phragmites australis* field scattered with *Alnus hirsuta* var. *sibirica* and *Salix* spp. The management was undertaken by 13 members, mainly on the weekends. The work done by these experienced members was very efficient. Within 2 months, the yatoda landscape had been revealed (Figure 5.11). Nothing was brought in from outside. Road reconstruction was done with thinned branches, *P. chino*, and vine from the local satoyama landscape woodland.

Fig. 5.11. An aerial photograph of Shinmei Yato

In the following year, the overall management and repair of the revived yatoda structure continued. Buried water reservoirs were excavated and wetland environments that would provide habitat for diverse species were revealed. The thinning of trees and the mowing of undergrowth in the coppice woodland were started that winter. The revived coppice was once again bright with sunshine, even in the understory, in early spring. The reservoir brimmed with water and rice paddies were in production. The entire yatoda regained its vitality and beauty (Figures 5.12, 5.13, 5.14, 5.15).

The management approaches, in general, were as follows:

1. Vegetation management: extraction of grass, fallen leaves, and damaged trees.
2. Yato management: recreation of the neglected yato and management in order to create a wetland habitat for diverse species.
3. Wildlife management: extermination of harmful species, such as crows in this case, in order to protect endangered species.
4. Usage management: patrolling, observing and cleaning.

5.3.3.2 Biodiversity Due to Restoration

Extirpated species gradually returned as the project progressed. The first species rediscovered was the Common kingfisher (*Alcedo atthis*). Kingfishers hunt loaches

Fig. 5.12. Before restoration of Shinmei Yato (1995)

Fig. 5.13. After restoration of Shinmei Yato (1997)

Fig. 5.14. Restoration of the irrigation pond by members of the Machida Historic Environmental Management Organization

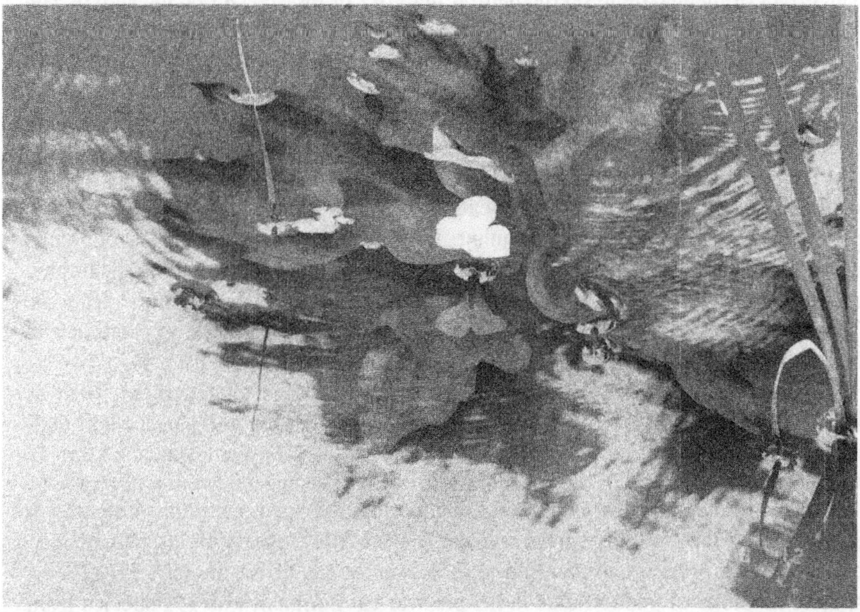

Fig. 5.15. *Ottelia alismoides* in bloom in a satoyama irrigation pond

in rice paddies. The Gray heron (*Ardea cinerea*) and Little egret (*Egretta Garazetta*) are now frequent visitors to the area as well.

Furthermore, species were seen through the seasons as the entire satoyama management progressed. In the bright coppice of springtime, colorful plants, such as *Viola grypoceras* and *Disporum smilacinum*, cover the ground and the sound of frogs can be heard from adjacent rice paddies. In early summer, the chirping of summer birds, such as the Narissus flycatcher (*Ficedula narcissina*) and Blue-and-white flycatcher (*Cyanoptila cyanomelana*), which are on a break from their long migrantion journey, echoes through the woodland. Dragonflies buzz about on midsummer days, and fireflies sparkle in the summer night. The Ural owl (*Strix uralensis*) hunts for rats. During the autumn harvest, each plant attracts birds and insects with its ripened fruit. The shrill voice of the Bull-headed shrike (*Lanius bucephalus*) breaks the stillness. A lucky visitor may encounter a mother and young Gray-faced buzzard (*Butastur indicus*) floating up through the sky with an ascending circular current. In winter, ducks toddle on frozen reservoirs. Flocks of the Great tit (*Parus major*), Varied tit (*Parus varius*), Long-tailed tit (*Aegithalos caudatus*), Japanese whiteeye (*Zesterops japonicus*), and Japanese pygmy woodpecker (*Dendorocops kizuki*) twitter and peck branches, seeking food. If you listen closely, footsteps of Dusky thrush (*Turdus naumanni*) and Siberian bluechat (*Tarsiger cyanurus*) walking on dryfallen leaves can be heard. Abundant wildlife reminds us of the energy and vitality of Mother Nature.

5.3.3.3 Paddy Field Vegetation after Restoration

The effect of management on the natural integrity of the site can be measured. The following are the results of an investigation of rice paddy vegetation conducted in the Shinmei Yato, the first place where the management operation was implemented.

The Shinmei Yato is one of the sources of the Tsurumi River system that originates in the north of Machida City. The Tsurumi River flows through the Tsurumi district, Yokohama, and pours into Tokyo Bay. The Shinmei Yato has a great volume of gushing water. After recreating the yatsuda structure in cultivated rice paddies between October and December, 1996, water inflow from the surroundings was used to make four permanent dry sections, five almost permanently saturated paddy sections, two permanent paddy sections, and two relatively moist sections. These paddies were planned to be maintained as diverse aquatic vegetation and hydrophyte species conservation habitats.

A total of 13 paddies were provided with appropriate management, such as leaving it untouched, regular mowing of grass, and plowing and surface soil paddling. During the period April to November, 1997, 119 plant species from 42 families were observed in these paddies. Of these, 115 species were native plants. This is good diversity for paddy land of less than 300 m^2 (excluding footpath). It even exceeds the diversity of the pre-war (before 1941) paddy fields, where agriculture was in progress. According to the observations of Kasahara (1951, 1954), species diversity of paddy fields in the pre-war environment was 191 species from 43 families nationwide. Of these, in the Higashi Yama area, including the Kanto

Table 5.3 Major species found in restored paddy fields in Shinmei Yato (April–November, 1997)

Family	Scientific name	Life form	Dormancy	Disseminule
Equisetaceae	*Equisetum arvense*	Perennial	G	D1
Typhaceae	*Typha latifolia*	Perennial	HH	D1
Typhaceae	*Typha orientalis*	Perennial	HH	D1
Gramineae	*Echinochloa oryzoides*	Annual	HH(Th)	D2, D4
Gramineae	*Arthraxon hispidus*	Annual	Th	D4
Gramineae	*Isachne globosa*	Perennial	HH	D4
Gramineae	*Leersia sayanuka*	Perennial	HH	D4
Gramineae	*Phragmites australis*	Perennial	HH	D1
Cyperaceae	*Carex dickinsii*	Perennial	HH	D4
Cyperaceae	*Carex maximowiczii*	Perennial	HH	D4
Cyperaceae	*Carex thunbergii*	Perennial	HH	D4
Cyperaceae	*Cyperus orthostachyus*	Annual	HH(Th)	D4
Cyperaceae	*Eleocharis wichurae*	Perennial	HH	D4
Cyperaceae	*Scirpus juncoides*	Annual	HH(Th)	D4
Commelinaceae	*Murdannia keisak*	Annual	HH(Th)	D4
Pontederiaceae	*Monochoria vaginalis* var. *plantaginea*	Annual	HH(Th)	D1, D4
Juncaceae	*Juncus effusus* var. *decipiens*	Perennial	HH	D4
Juncaceae	*Juncus leschenaultii*	Perennial	HH	D4
Saururaceae	*Houttuynia cordata*	Perennial	G	D4
Polygonaceae	*Persicaria nipponensis*	Annual	HH	D4
Polygonaceae	*Persicaria thunbergii*	Annual	HH(Th)	D4
Ranunculaceae	*Ranunculus cantoniensis*	Perennial	HH	D2, D4
Guttiferae	*Sarothra laxa*	Annual –perennial	HH	D4
Oenotheraceae	*Epilobium pyrricholophum*	Perennial	HH	D4
Oenotheraceae	*Ludwigia epilobioides*	Annual	HH	D4
Umbelliferae	*Oenanthe javanica*	Perennial	HH	D4
Primulaceae	*Lysimachia fortunei*	Perennial	HH	D4
Labiatae	*Lycopus ramosissimus* var. *japonicus*	Perennial	HH	D4
Scrophulariaceae	*Lindernia procumbens*	Annual	HH (Th)	D4
Compositae	*Bidens frondosa*	Annual	Th	D1, D2

Th, annual; G, geophyte; HH, aquatic plant or hydrophytes; D1, anemochore or hydrochore; D2, zoochore; D4, clithochore.
Paddy weeds were included in the HH form.

Table 5.4 Species variation in response to soil moisture regimens and agricultural management: species that appeared between April and November, 1997

Soil moisture	Flood	Saturated	Saturated	Hydric	Mesic
Management	Early spring plow	Non-management	Mowing twice/year	Mowing twice/year	Mowing twice/year
Average no. species	29	23	28	58	36
Average no. paddy weeds	21	13	17	29	19
Ratio of paddy weeds to total species	72%	57%	61%	50%	53%
Average no. species after management	15	10	16	56	13
Vegetation cover, top 5	*Monochoria vaginalis* var. *plantagine* *Murdannia keisak* *Echinochloa oryzoides* *Leersia sayanuka* *Limnophilia sessiliflora*	*Phragmites australis* *Leersia sayanuka* *Persicaria thunbergii*	*Phragmites australis* *Persicaria thunbergii* *Leersia sayanuka* *Oenanthe javanica* *Isachne globosa*	*Equisetum arvense* *Isachne globosa* *Houttuyni cordata* *Artemisia princeps* *Oenanthe javanica*	*Equisetum arvens* *Houttuynia cordata* *Glechoma hederacea* ssp. *grandis* *Persicaria thunbergii* *Isachne globosa*
Overview of vegetation	Mostly paddy weeds	Dominated by *P. australis,* with a few other species	*P. australis* is 60 cm in height Species are more diverse than at an unmanaged lot	Various species (including non-wetland species)	Most species are common in non-wetlands

Agricultural management proceeded in summer.

area, there were 182 species from 42 families, while in the Shinmei Yato thre were 68 species from 21 families. Yatoda landscapes, where big machines cannot be applied and traditional cultivation was carried on for generations, have been sanctuaries for paddy plants for a quarter of a century (Yamada et al. 2000).

Species investigations in 1997 observed important plant species designated in the Red Data Wildlife and Plants in Tokyo Metropolis (Bureau of Environmental Protection, Tokyo Metropolis 1998). *Callitriche verna* and *Dopatrium junceum* (A rank: distinction endangered species), *Limnophila sessiliflora*, *Alisma canaliculatum*, *Pilea hamaoi*, *Glyceria autiflora*, *Eleocharis wichurae*, and *Acorus calamus* var. *angustatus* (C rank: species for which habitat deterioration may result in them becoming rank A or B) are all included. A 1998 investigation recorded *Isoetes japonica* and *Elatine trianda* var. *podicellata* as A rank species.

With a total of 13 sections, species that were observed in more than seven sections are shown in Table 5.3. Other than *Equisetum arvense* and *Houttuynia cordata*, the rest are all paddy field weeds (in the sense that anything other than rice can be considered a weed). Each of the 13 sections differs slightly in area, sun exposure, and hydrology due to slight topographic differences. Surrounding woodlands and banks may supply different species seeds. The average number of species observed in each section was 20. Comparison of soil moisture and management (Table 5.4) indicates that species diversity is highest when grass is mowed twice a year, in June and August, and the growth of tall herbaceous plants, such as *P. australis*, is prevented. This is a consequence of the many species that inhabit the different hydrologic conditions, ranging from moist to dry, that appeared in this section. Before management was initiated, the entire section was covered with *P. australis* and *E. arvense*, but, after the management was complete in autumn, species distribution clearly followed the water conditions of the section (Table 5.4).

5.3.3.4 Restoration of Fauna

Recreation of the yato environment brought fauna back as well. Fukuyama and Sato (1998) conducted a frog investigation. Native amphibia reported during the winter of 1997, when the reconstruction was finished, until August 1998 were: Japanese fire-bellied newt (*Cynops pyrrhogaster*), Eastern-Japanese common toad (*Bufo japonicus formosus*), Japanese tree frog (*Hyla japonica*), Japanese brown frog (*Rana japonica*), Montane brown frog (*Rana ornativentris*), Tokyo daruma pond frog (*Rana porosa*), Wrinkled frog (*Rana rugosa*), and Schlegel's green tree frog (*Rhacophorus schlegelii*). The amphibious species distributed in the Kanto Plain included two Urodela and eight Anura, except for non-natives. Considering *Hynobis tokyoensis* and *Rhacophorus arbareus*, which are native to the area but were not recorded in the recreated yato, being mountainous species, it could be concluded that all amphibious species that are expected to be distributed in the area have been observed.

According to an investigation by Masami Hasegawa in 1997 (private letter) in the recreated Shinmei Yato paddy section, there were 11 egg masses of *R. japonica* and one egg mass of *R. ornativentris* observed. On the other hand, investigations conducted by Fukuyama 1 year later in 1998 recorded 1004 egg masses of *R. japonica* and 96 egg masses of *R. ornativentris* (Fukuyama and Sato 1998). Both species almost 100-fold in 1 year. Fukuyama concludes that it is likely frogs from surrounding environments came to lay eggs, but this itself is the consequence of the appearance of open waterfront due to recreation of the paddy environment (Fukuyama and Sato 1998). *Rhacophorus schlegelii*, which lives in trees, requires woodland in order to be able to move. The diversity of the frog species that require both a waterfront and land habitat could be an indicator of the natural integrity of the area composed of rice paddies, water channels, and water reservoirs with adequate slope, surrounded by layers of blocks of grassed bank and coppice woodlands.

Table 5.5 Frequency of dragonflies and damselflies found in Shinmei Yato

	Year and date					
	1997					
	5 May	26 May	19 Jun	9 July	25 July	29 Aug
Ischnura asiatika					△	
Cercion sieboldii						△
Lestes temporalis						
Asiagomphus melacnops		○	△			
Anotogaster sieboldii					△	△
Anax nigrofasciatus nigrofasciatus	△	○	△		△	△
A. nigrofasciatus nigrofasciatus					△	△
Epophthalmia elegans				△		
Ortheterum japonicum japonicum	○	□	○			
Ortheterum albistylum speciosum	△	△	△	○	□	□
Ortheterum triengulare melania		△	○	○	○	○
Crocothemis servilia mariannae				△	△	△
Sympetrum darwinianum						
Sympetrum frequens				△		
Sympetrum eroticum eroticum				△		△
Sympetrum parvulum				△		△
Sympetrum infuskatum						
Pantala flavescens				○	○	△
Rhyothemis fuliginosa						△

□, more than 11times; ○, six to 10 times; △, one to five times.

In the case of insects, Ogino (1998) investigated dragonfly species. Seventeen dragonfly species were recorded between May, 1997, and June, 1998, in Shinmei Yato (Table 5.5). These were species commonly found in yato environments, although the increasing diversity is significant considering that there were only six species recorded before management was started. Regular mowing was found to

Year and date										
		1997						1998		
30 Sep	30 Sep	26 Oct	24 Nov	20 Dec	7 Feb	3 Mar	29 Mar	2 May	27 May	20 Jun
		△								
								△	△	△
								△	△	△
								□	○	△
○	○	△						△	△	○
									△	△
									△	△
△	△	△								
□	□	□	○	△						
△	△	○	△							
○	○	○	△							
△	△	△								
○	○									

result in prolific shrubby wetland-loving species. Three Red Data species were recorded: ***Rhyothemis fuliginosa*** is designated as rank B, whereas ***Asiagomphus melaenops*** and ***Sympetrum parvulum*** are rank C species in the Red Data Wildlife and Plants in Tokyo Metropolis (Bureau of Environmental Protection, Tokyo Metropolis 1998).

5.3.4 The Future of Traditional Satoyama Management Procedures

The example of the Machida Historical Environment Management Organization illustrates the development of a system to continue adequate management of satoyama landscapes. The land is considered public property with a conservation designation and is managed using traditional methods by local residents.

Traditional procedures for the management of satoyama landscapes, inherited over generations, may be effective in conserving species diversity that was also created over time, as is apparent in the great species diversity seen in the recreated yato (Figure 5.16).

There is no doubt that the procedure is effective in suburban satoyama conservation. There is no reason that it should not be introduced to other areas as well. However, some problems remain. In suburbia, where farmers are lacking heirs, who is going to inherit the procedure? In addition, who is going to undertake the management? Who is going to provide sponsorship when government budgets are tight? Individual management systems by local residents will need to be investigated, such as inheriting the traditional management procedure systematically, fundraising with National Trust, and so forth.

Fig. 5.16. The managed coppice

5.4 Restoration of Aquatic Ecosystems

I. Washitani

5.4.1 Restoring Lakes with *Soda*

If you can identify the contemporary value in the biological resource of satoyama landscapes, there might be a possibility of restoring its economic viability and maintaining it (Washitani 2001b). The Asaza Project in the Kasumigaura area may be highlighted as one such example. This project aimed to restore the aquatic ecosystem by using *soda* brought from the coppice surrounding the lake (Washitani & Iijima 1999). This use is having an influence on the conservation of coppice woodland biodiversity. Soda consists of bundled twigs and branches from coppice trees. A cooperative to make this soda was set up, which not only helped the environment, but also created jobs. In this section, the collaboration between the Asaza Project and the Soda Cooperative is explained by referring to the background and work these groups have been doing and what is envisaged for the future of these two units.

Kasumigaura Lake is the second-largest lake in Japan, but it has an average depth of only 4 m. Kasumigaura Lake is located downstream of major rivers, so nutrients accumulate readily in the lake. This means that the biological productivity is relatively high, as shown by fisheries that have been there since the Nara period (for more than 1000 years). During the 1970s, the water quality deteriorated because of urban development taking place in the watershed of the lake. Although many pilot trials were attempted to improve the water quality, they did not have much effect. It is now understood that the reason why water quality is not improving is because of the *improvements* made to the lake: concrete dykes and the consequent loss of purifying vegetation.

The lake was originally characterized by ecotones (i.e., in this case, a transition section between the water body and the land, between the aquatic and terrestrial ecosystems). Because the slope is very gentle, many different types of plants grew there because of the surface substrate and the shallow water level. These areas acted not only as gateways of energy, materials, and wildlife from aquatic to terrestrial ecosystems and vice versa, but also as zones that controlled the inflow and outflow of materials. The loss of this ecotone due to the concrete walls resulted in a cessation of this interaction and brought about the deterioration of the purifying ability and an overwhelming accumulation of nutrients. This phenomenon can be seen in many lakes around the world.

Once the ecotone in a lake is lost, even with high-technology water purifying technology, it is very difficult to improve the water quality. Changes in the plankton community bring about explosive eutrophication. As a result, the water becomes bad smelling and may affect human health. Since the early 1970s, many projects have attempted to improve water quality, including regulations to limit sewage etc.; so far, water quality has not improved and has only become worse. The Asaza

Fig. 5.17. A *Nymphoides peltata* community in Kasumigaura Lake

Project aimed to improve the health of the lake by restoring the vegetation in the ecotone (Figure 5.17).

5.4.2 Asaza as a Symbol

Asaza (*Nymphoides peltata*) is a water lily that can be found in most of Eurasia. In Japan, as a whole, it is one of the most common species in lakes, ponds, and water channels Figure 5.18). Today, because of the concrete walls and the water quality problems, almost one-third of aquatic species are on the species red list, and Asaza is among these species as a class 2 species. Kasumigaura Lake used to have the largest Asaza community in Japan, but there has been a significant decline in Asaza in Kasumigaura Lake because of the concrete walls.

Two types of Asaza flowers are found. To breed, both types are needed. Only a single type is now found in Kasumigaura Lake. For healthy plants, interaction is necessary. Asaza communities with both types of flowers have decreased significantly, so no succession is taking place.

The flower of the Asaza blooms between summer and autumn, and the seeds float on the water and will wash onto the shore. The seeds do not germinate immediately, but wait until spring. The seeds must wash onto the shore to start to grow; this process is also hampered by the concrete walls.

Under the monsoon climate, the water level changes in the lake had a connection with the lifecycle of the plant. The Asaza seeds germinate on the ground and the water level then rises, the leaves float on top of the water during the monsoon cycle

Fig. 5.18. Flowers of *Nymphoides peltata*

and the water level changes. Where the concrete walls surround the shore, these changes cannot take place, because the bank is too steep. Areas are either above or below the water all the time.

Places that contain Asaza are those where other aquatic plants can be found, namely on the few remaining, although once abundant, sandy beaches.

However, exposed places above water are often covered with Fragmite; this means that, in summer, Fragmite dominates the area, blocking the sunshine for the Asaza. Even though places are suitable for Asaza because of the landform, the Fragmite dominates. This situation means that it is hopeless for the Asaza to regenerate in the future. In no conservation precautions are taken, quite soon Azasa will become extinct (Washitani 1994).

Since the autumn of 2000, an operation to lower the water level during the winter has taken place to restore, to a certain extent, the habitat of aquatic plants.

5.4.3 Collaboration Proposed by the Asaza Project

In 1995, in Kasumigaura Lake, the concreting of the walls was finished and the control of the water level was finalized; this coincided with a drop in the number of

Azasa. In addition, research by scientists confirmed the Asaza lifecycle. A research group composed of the Japanese Ministry of Construction (now Ministry of Land, Infrastructure and Transport), an NPO and the author (Izumi Washitani), conducted a survey and showed that the rapid decrease of the Asaza community started with the control of the water level. Within a couple of years, the Asaza community had decreased to 10% of its former level. Also, a large number of changes in the aquatic ecosystem were observed because of the construction. Since the mid-1990s, the fisheries have suffered a decrease in the smelt harvest, the traditional catch of the area. Only introduced fishes have come to be caught in the lake.

To remedy this serious situation, the Asaza Project was established in 1995, as a partnership between the Ministry of Construction, citizens, schools, and local volunteers. This project worked on restoring the Asaza community in front of the concrete walls. The target of this is to soften the wave movement to promote the laying of sand and create an ideal environment for Asaza to grow. This will then promote not only Asaza, but also the whole aquatic plant community in Kasumigaura Lake, which may act as a nursery bed for native fishes.

In terms of nutrient balance, the issue is complex, so it is not easy to tell what system is working there. The Asaza Project group expects that the restoration of the plant community may help the nutrient balance in the lake too. The floating leaves have a high photosynthetic ability and an ability to capture phosphates and nitrates. This means they will clean up the water. This lake is rich in nutrients; if the Asaza have a suitable habitat, they may grow rapidly.

There are also insects and birds that feed on Asaza leaves, which means that the nutrients captured by the Asaza will be brought out of the lake. Also, most of the aquatic plants, such as Asaza, have a system of bringing oxygen to their roots. The environment surrounding the roots becomes relatively high in oxygen, promoting certain bacteria that denitrify the water. The water is also improved by capturing the floating substance in the plants, further improving quality.

The current situation in Kasumigaura Lake is that even though the Asaza Project group has tried to improve the land area, the number of Asaza is limited and the seeds have difficulty in flourishing. The Asaza Project group has a project known as *Adopt*, in which citizens grow seedlings in their aquarium or backyard to germinate the seeds, which are then transplanted to the lake. Because this collaborative project between the government and citizens came to be one of the leading projects in Japan, it received the Asahi Newspapers' Environment leadership award in March 2000.

5.4.4 The Adopt Program

The Adopt Program was started in 1995, but that year saw the transplanted Asaza all being washed away. The Asaza Project group realized that the waves of the lake were the problem. Soda was something that was used as a traditional method for dampening the waves. These bundles of twigs and branches brought from the satoyama coppices not only ease the impact of the waves, but can also be a place for fish to use as a habitat. Being natural, the soda will not deteriorate the water quality of the lake. If the sand movement can be stopped using soda, the planting of the Asaza can take place. The Asaza Project used soda from the surrounding satoyama landscape.

The Japanese Ministry of Construction is implementing a major project to promote this program, requiring a huge amount of soda. This considerable need means that the satoyama coppices will be maintained. Unfortunately, the practice of using soda has been abandoned for so long that the techniques and land use no longer exist. The project had to form its own cooperative for to produce the soda. The cooperation will result in the maintenance of the coppice woodlands and the conservation of biodiversity in the whole area. The surrounding satoyama landscape has been abandoned for a long time, with the woodlands dominated by *Pleioblastus chino* brought about by the loss of economic viability, and a consequent loss in biodiversity. Hopefully the demand for soda will improve the situation. The Kasumigaura Lake Soda Cooperative was established in 1999 to provide soda for restoration of the whole lake. The aim of the cooperative is to make money and also to improve the biodiversity of the whole area and the lake. This cooperative can be seen as a new type of business, which combines profitability and environmental conservation. The Soda Cooperative has expanded the area maintained by 100 ha per year. Not only are young people involved with starting the new business, but so are the recently unemployed. It is hoped that this cooperative may improve the satoyama landscape situation, which, until now, has only been discussed in terms of government or volunteer projects.

The maintenance and harvesting of coppice woodlands operated through effective resource use, but with an environmental perspective on the way in which soda is harvested. The fundamental attitude of the Soda Cooperative toward the coppice woodlands is not only economic, but also ecological.

6 Biological Resources

6.1 Biological Resources in Village Life

K. Iiyama

6.1.1 Satoyama Landscapes in Folklore

Many Japanese folktales begin with words such as:

> A long, long time ago there lived an old man and his wife. The man cut firewood in the mountains and his wife washed clothes in the river.

Villagers earned their living by gathering firewood in the mountains and making charcoal to sell to the towns. This can also be seen in other folktales such as the following:

> A long time ago, a charcoal burner called Santaro lived alone in the mountain village of Minamitani in Takao. He was a hard worker and burned good charcoal diligently using a charcoal burner he built himself.

In addition to this, the mountain as the place for hunting and the village as the place of food production were organically combined and, as a whole, represented the place of production. This situation can be seen in the folktale of *The Meeting of a Fox and a Raccoon Dog in a Rice Field*:

> A fox and a raccoon dog lived at a certain place a long time ago. One fine day they met, and the fox began, 'Well,' he said, 'human beings cultivate fields like this, and produce a variety of foods. We could also do the same thing.'
>
> 'Indeed, this is correct,' replied the raccoon dog, 'such a thing could be considered. I cannot hunt enough game even if I walk for days around the mountain searching for food. Human beings, though, are quite clever. They plunder the mountain and freely hunt game to eat with a rifle so that only a little game is available to us. In that case, we have no choice but to break the soil and till a field as they do.'
>
> 'But, we need a tool called a pickaxe,' said the fox.
>
> The raccoon dog replied, 'Well, I will take the shape of a human being and copy the right tools.' They then copied pickaxes and plowed a

field. A fine field was prepared and they said, 'Let us plant this field (Folktale of Okayama, Web.).'

These long opening passages from Japanese folktales show how important such places can be to the Japanese. The satoyama landscape life is a poor existence, despite the riches of nature and the sweat of people's brows, and is vividly depicted in many folktales handed down over countless generations.

6.1.2 Sustainable Production of Biological Resources

Satoyama environments, which produce various biological resources indispensable to life, have not only been of economic benefit, but have also permitted human survival in Japan, where the terrain is mountainous and the features intricate. Coppiced woodlands supply wood for building materials, furniture, tools, firewood, charcoal, and game, as well as foods such as wild herbs and fruits. In addition, they supply resins used as medicinal herbs or varnishes. Satoyama grasslands were used for grazing oxen and horses. Fields were used for food production and the cool river running through the fields was the place for fishing and for obtaining fresh water. Because satoyama landscapes provided abundant organic matter for use as fertilizers, people would have known from experience that they could obtain not only a varied, but also a rich, harvest in these areas. Furthermore, in folktales that include animals such as foxes, raccoon dogs, apes, rats, snakes, sparrows, pheasants, cranes, woodpeckers, and freshwater snails, satoyama environments are said to be places of production and habitats for a variety of life.

Satoyama environments, providing firewood, charcoal, building materials, fallen leaves, pasture for oxen and horses, and, occasionally, stone, as well as rice and other foods, formed the basis of farmers' livelihoods before the modern economic age and the distribution revolution. Farmers formed a voluntary local community that maintained satoyama environments with care, selflessness, wisdom, and labor.

In almost all cases, satoyama environments did not belong to the people living and using the land, but were made up of public land or of a landowner's large, private holding. People living in satoyama environments formed a local community to manage the landscape voluntarily so as to obtain their staple foods. On the other hand, however, the landowners recognized the importance of good management to maintain the satoyama landscapes. As a result, the local community transformed into an *iriai-shudan*, or a common group, the members of which consisted of people able to exercise their management rights on a coppice woodland, a wilderness, and other lands. The right came to be recognized as *iriai-ken*, or a common right, notwithstanding a feudal and totalitarian society. *Iriai-ken* is a right to manage the land, for regulating *iriai-shudan* and the harvest. This right was conserved as part of the Civil Code under the rule of law following the Meiji Revolution. The biological production of satoyama environments was maintained until the economic value of such areas collapsed as a result of the

1960s fuel revolution and the rapid decline of agriculture and forestry, brought about by the high level of economic growth.

6.1.3 Coppice Woodlands and Biomass Productivity

Satoyama landscapes formed the basis for self-sufficiency under the economic regimen at the time by allowing surplus products and cottage industry goods to be taken to an exchange market. Therefore, the number of inhabitants that satoyama landscapes could accommodate depended on the amount of biomass production. The following is an estimation of the biomass produced by a satoyama landscape and how many inhabitants it could support.

The Food Agency of Japan estimated, in 2000, that rice consumption in Japan is 65 kg per person per year. The productivity of a rice paddy can be estimated based on 1 bag (60 kg) per person per year, even though rice consumption is decreasing. Although the quantity of rice produced is 5.0 t per ha (2000) on a cleaned-rice basis, the mean quantity of production for the 5 years between 1878 and 1882, for which there are existing statistics, was 1.76 t per ha in unpolished rice, equivalent to 1.53 t per ha on a cleaned-rice basis (Agricultural Policy Research Committee 1969). In that case, a 1-ha paddy field was able to support 25

Table 6.1 Annual household fuel consumption, according to fuel type, in urban and rural areas, 1995 and 1989

Area	Electricity		LPG		Kerosene		Fuel wood		Charcoal		Biomass residue	
	GWH		10^3 tons		m^3		10^3 tons		10^3 tons		10^3 tons	
	1995	1989	1995	1989	1995	1989	1995	1989	1995	1989	1995	1989
Philippines	8134	6845	503	321	776	496	14 557	18 317	770	1568	3668	2570
(%)	28.6	22.7	1.8	1.1	2.7	1.6	51.2	60.8	2.7	5.2	12.9	8.5
Urban	6456	5052	384	243	397	215	3886	3258	386	837	1003	433
(%)	51.6	50.3	3.1	2.4	3.2	2.1	31.1	32.5	3.1	8.3	8.0	4.3
Rural	1678	1793	119	78	379	281	10 671	15 058	384	728	2665	2138
(%)	10.6	8.9	0.7	0.4	2.4	1.4	67.1	75.0	2.4	3.6	16.8	10.6
NCR	3191	2867	150	133	128	83	54	132	28	121	13	19
(%)	89.5	85.5	4.2	4.0	3.6	2.5	1.5	3.9	0.8	3.6	0.4	0.6

GWH, giga watts (billion) – 10^9watts; LPG, liquefied petroleum gas; NCR, national capital region.
Source: National Statistic Office, Department of Energy, Republic of the Philippines (1989, 1995).

Table 6.2 Annual growth and CO_2 fixation by trees

Classification		Annual growth			Accumulation		Annual increase (%)
		Carbon t/ha/year	CO_2 t/ha/year	Biomass t/ha/year	Volume m^3/ha	Weight t/ha	
Japanese trees	Japanese cedar	0.7-1.4	2.5-5.0	1.7-3.4	120	48	3.5-7.1
Tropical trees	Shorea*1	4.5	16.5	11.3	180	126	8.9
	Mahogany*2	3.0	11.0	7.5	180	108	6.9
	Narra*2	4.0	14.7	10.0	180	108	9.3
Melaleuca	Swamp dried up at dry season*3	5.7	20.9	14.3	180	144	9.9
	Never dry swamp*3	10.0	36.7	25.0	180	144	17.4

Japanese cedar, *Cryptomeria japonica*; Shorea, *Shorea curtisii*; Mahogany, *Swietenia macrophylla*; Narra, *Pterocarpus indicus*; Melaleuca, *Melaleuca cajuputi.*
*1Determined by Tange et al. 2000 in Malaysia.
*2Determined by Ono et al. 2002 at Los Banos, Philippines.
*3Determined by Yamanoshita et al. 1999 at Narathiwat, Thailand.

people or five families, assuming five members per family.

Next, let us consider the biomass productivity of satoyama landscapes in terms of fuel supply. Unfortunately, statistics regarding annual per capita consumption of biomass fuel cannot be found. Although it is difficult to compare directly because climate conditions are completely different, the consumption of biomass fuel can be estimated based on data from the Philippines. The population of the Philippines was 73 000 000 in 1998 and 31 500 000 (43.3% of the total population) resided in rural areas (FAOSTAT homepage). In the rural areas of the Philippines, 86.3% of energy consumption was from biomass fuels (the sum total of firewood, charcoal, and other biomass resources is shown in Table 6.1;

Table 6.3 Annual production of tropical forage grasses (dry weight; t per ha)

Foragegrass	Year				
	1984	1985	1986	1987	Average
Rhodes grass	15.5	21.1	16.8	18.3	17.9
Guinea grass	10.9	17.3	16.8	22.4	16.9
Green panic	11.8	16.2	15.9	19.8	15.9
Setaria grass	12.8	24.3	24.0	28.0	22.3
Buffel grass	9.9	15.0	14.7	16.7	14.1

Modified from Shinjo and Hoshino 1989.
Rhodes grass, *Chloris gayana* cv. *katambora*; Guinea grass, *Panicum maximum* cv. *gatton panic*; Green panic, *Panicum maximum* var. *trichoglume Robijns*; Setaria grass, *Setaria sphacelate* cv. *kazungula*; Buffel grass, *Cenchrus ciliaris* cv. *biloela*.

National Statistics Office, Department of Energy, Republic of the Philippines 1989, 1995).

Therefore, the annual per capita consumption of biomass fuel is estimated to be 0.5 t (1 m^3, assuming the specific gravity of biomass resources is 0.5). Because the calorific value of 1 kg firewood is 4500 kcal (approximately 19 000 J), the annual per capita consumption of biomass fuel corresponds to 1.37 kg firewood or 6100 kcal (approximately 26 000 J).

The average amount of biomass accumulation in a Japanese coppice is 120 m^3 per ha, with an average annual growth of 5% (6 m^3 per ha), so the feasible amount of biomass supply for sustainable production is 4 m^3 per ha (2 t per ha; Table 6.2). Therefore, a 1-ha coppice could support four people. A 1.2-ha coppice would be required to support one family. This only takes fuel use into account and a 1.5-ha coppice would also be required per family when considering the cottage industry goods for cash earnings.

The cow and the horse were indispensable for farm work. These animals consume 8 kg grass in dry weight (2 kg assuming the rate of digestion as 25%) daily or approximately 3 t annually. The annual grass growth of a well-managed pasture has been measured at 17–22 t per ha in dry weight in Iriomote Island, Okinawa, in the subtropical zones (Table 6.3). An annual growth of 12 t per ha was measured in the experimental farm of the Graduate School of Agriculture and Life Sciences, the University of Tokyo in Bunkyo-ku, Tokyo (Ono et al.2002). With these results, the annual grass growth in temperate grassland was estimated to be 8–12 t per ha. Therefore, three to four cows or horses could be kept on 1 ha grassland. This means that a 4-ha paddy field, a 30-ha coppice, and a 2–3-ha grassland were the minimum requirements for the formation of a community consisting of 20 families (a population of 100, with a cow or horse for every three families).

Table 6.4 Example of habitat in a small satoyama on the outskirts of an urban area

Classification of habitat	Distribution (%)	Area (ha)	Storage (m^3/ha)	Storage (t/ha)	Total storage (t)	Annual increase in storage (t/year)	Available quantity for effective utilization (t/year)
Non-coniferous	60	18	120	60	1080	54	40
Coniferous	20	6	150	75	450	23	17
Grassland	10	3		10	30	30	22
Others	10	3		5	15	15	11
Total	100	30			1575	122	90

Table 6.5 Example of habitat in a large satoyama

Classification of habitat	Distribution (%)	Area (ha)	Storage (m^3/ha)	Storage (t/ha)	Total storage (t)	Annual increase in storage (t/year)	Available quantity for effective utilization (t/year)
Non-coniferous	43	215	120	60	12,900	645	484
Coniferous	46	230	150	75	17,250	863	647
Grassland	10	50		10	500	500	375
Others	1	5		5	25	25	19
Total	100	500			30,675	2,033	1,525

Storage and annual increase in storage were estimated from Tables 6.2 and 6.3.

The size of satoyama landscapes varies with geography. The overview of biomass production and utilization given here assumes two types of satoyama landscape: small (30 ha) satoyama landscapes in the city suburbs (Table 6.4); and large satoyama landscapes (500 ha) (Table 6.5), consisting of secondary woods, afforested areas, and grasslands, with several surrounding satoyama environments.

Let us assume, in the small-type of satoyama landscape, that the traditional rural landscape remains. The patch areas are small and include fields on the slopes of hills and paddy fields forming intricate *yatsu*, evergreen broad-leaved forests and coniferous–deciduous coppice woodland patches. These contribute to the typical satoyama landscape of city suburbs. The vegetation distribution of this type of satoyama landscape is shown in Table 6.4, suggesting that this satoyama landscape could accommodate only 20 self-sufficient families. The wood available for durable biomass production is only 90 t (200 m^3).

On the other hand, the vegetation distribution in the large-type of satoyama landscape is shown in Table 6.5, including secondary woods (215 ha) of broad-leaved trees, such as *Quercus serrata* and *Quercus variabilis*, and relatively young coniferous trees planted as forestry crops. The available wood resource in durable biomass production in this satoyama landscape is estimated to be approximately 1525 t (3500 m^3).

6.1.4 Potential for Use of Satoyama Woody Biomass

The Japanese paper industry consumes 18 800 000 t woody biomass annually as raw material (Japan Paper Association homepage). One mill in particular consumes 640 000 t annually (1800 t daily; Tomakomai Mill, Oji Paper Co. Ltd homepage). The available annual wood resource of the above-mentioned large satoyama landscape would be used in a single day by the plant. Moreover, all the satoyama landscape wood would be used up in 1 month of factory operation. As well as paper, 6 000 000 t wood annually, or 16 000 t daily, is used in Japan to manufacture goods such as plywood (3 260 000 t per year), particleboards (1 080 000 t per year), and fiberboards (1 800 000 t per year).

Compared with these figures, it is easy to see that the 1525 t wood produced annually by the large satoyama landscapes is small, let alone the 90 t produced by the small satoyama landscapes. Although the quantity distributed in all areas of satoyama landscapes would be large if it could be collected and totalled, the transportation cost of such an operation would be prohibitive. Therefore, satoyama landscape wood cannot supply raw materials for the wood-consuming industry at present and cannot compete with the output from the natural mountainous regions, distant from human settlements, which are maintained for supplying coppice wood.

Furthermore, comparisons are difficult between satoyama landscapes because they have different scales and regional characteristics. Therefore, it is necessary to examine individually the use of resources according to the scale and regional characteristics. So, let us examine in detail the small and large types of satoyama landscapes.

It can be expected that small-scale satoyama landscapes located in the city suburbs will function as places of relaxation for urban dwellers for their recreation and for practicing and experimenting in environmental production and other activities. Although many areas of satoyama landscapes in the city suburbs are currently used in such a manner, they should be kept in better condition for such purposes. Farms and studios for rent, with adequate accommodation and resting places, should be installed and the raw materials from satoyama landscapes should be used for the manufacture of handicrafts, charcoal burning, and so forth. Satoyama landscapes could also be used as places where people can experience environmental restoration and ecological conservation and restoration. This might include the utilization of charcoal made under the supervision of trained instructors for an environmental restoration project, for soil improvement, and the like.

The amount of resources available from large-scale satoyama landscapes was estimated to be 1525 t per year. Although this may seem like a high figure, secondary growth, especially that of secondary woodlands, cannot be conserved unless resources are consumed at this rate every year. The ideal growth of stands in forestry plantations, mainly coniferous trees, which make up the majority of secondary woods, can only be realized by moderate thinning of trees. The same can be said for the ideal growth, germination and renewal of regenerative coppices of secondary woodlands that consist mainly of angiosperm trees. If moderate clearing is not carried out, the canopy becomes closed by high dense trees that obstruct the light. The germination of plants on the floor of the woodland would be restricted, resulting in a loss of vegetative diversity. The habitat and breeding of raptors, such as hawks, which feed on other birds and small animals on the forest floor woodland, will also be disrupted because the small animals will not be able to be seen from the sky. Thus, in order to conserve woodlands under sound management and to enhance the economy of the area, the available wood resources of satoyama landscapes must be utilized effectively and in a different manner to that which exists under the present management regimen.

6.1.5 Technology Employed to Use Woody Biomass

Although logs of superior quality may also come out of artificial plantations in large-scale satoyama landscapes, most coppice wood supplied from satoyama landscapes, where the production of woody biomass is not the main purpose, is the result of thinning and is, therefore, regarded as ordinary or of low quality. If the market for ordinary or low-quality wood materials grew, the thinning out of trees would occur without the need for added incentives. However, these materials cannot be sold unless they have been processed to some extent into goods such as laminated timbers or lumber. Satoyama landscape owners cannot make a return on their investment without undertaking such processing and the regional economy would soon become exhausted. The development of effective utilization of woody biomass from satoyama landscapes, in addition to manufacturing laminated timbers and lumber, is becoming increasingly necessary in the local community.

What practical uses can be considered? The Japanese Research Association for Mining and Manufacturing Technology Law (1961) aimed at the establishment of a corporation to enable the national support of research and development of private enterprises, based on a cooperation among industries, institutions, and the government. The Association for Promoting Advanced Science and Technology of Forest Resource Utilization (APAST) was established under the Japan Forestry Agency and first carried out research and development to enable a high-technology utilization based on criteria set at the national level. For every specific trial research project, private enterprises cooperated with a technical research association. These reports propose, for example, the manufacture of composite panels made from wood fiber with charcoal, cement, and ceramics to provide materials for electromagnetic wave shielding. These technologies would be expected to replace plastics made from petroleum and adhesives by wood and adhesives made by liquefying the coppice wood. Moreover, the reports documented the development of a carbonization furnace that substitutes for the traditional kiln used for the production of charcoal. The manufactured charcoal would be used as soil-improvement material, a growth-promoting material for plants and for the regulation of soil microbe flora. In addition, equipment for efficiently collecting wood vinegar is detailed.

Steam-explosion treatment of timber was developed independently in Canada and Australia in the early 1960s. During steam-explosion treatment, raw wood materials are processed into fiber at high temperatures and pressures (200–220°C, 1–3 MPa) followed by a sudden reduction of pressure to ambient pressure. Because wood is a porous polymer material, it is fiberized when the high-temperature, high-pressure steam in the minute holes expands by instantaneous decompression. In addition to the physical destruction of the woody material, a chemical change results from the organic acids being generated from the woody material, through the use of high-temperature steam. This technique was originally developed to separate and isolate cellulose, hemicellulose, and lignin, an aromatic polymer, and to use these as resources in the chemical industry.

Glucose, xylose, and other products can be prepared from the separated cellulose and hemicellulose, and even ethanol can be manufactured by fermentation. Xylitol, currently used in chewing gum and other products as a sweetener that prevents cavities, can also be manufactured from xylose. Various products currently manufactured from petroleum would be substituted by chemicals from lignin. However, the process has not yet been commercialized due to high costs. Several researchers have succeeded in manufacturing lumber, which cleared the Japanese Agricultural Standard (JAS) of the highest quality and was surpassed only by steam-exploded pulp, by adjusting the explosion conditions on oil-palm frond woody biomass, without using any adhesives (Suzuki et al. 1998). This method would also be applicable to bamboo and various non-coniferous woods.

The demand for water-absorbing materials to be used in disposable diapers and other products has increased. The water-absorbing materials are manufactured by introducing carboxyl groups as the water-attracting groups to a synthetic polymer from petroleum or to cellulose as a natural polymer. Although the water-retention capacity of these water absorbent materials increases when the carboxyl group is introduced in great quantities, they also dissolve in water under the same conditions. For this reason, high polymers with a high quantity of carboxyl groups are bridged (known as cross-linking), using a chemical reagent, to be insoluble in water.

In the case of coniferous woods, insoluble water absorbent materials have been obtained by carboxymethylation of the steam-exploded pulp without any foreign agents (Lam et al. 1985). This has been processed chemically and a material that absorbs heavy metals selectively and efficiently has been developed (Lam et al.

1986). These trials have aimed to demonstrate the use of such materials to hold water in desert soils, for example, and to treat water polluted with heavy metals.

In future, converting renewable coppice wood into energy and using it as a substitute for fossil fuels is expected to contribute to the conservation of a living environment as a clean energy resource that emits few air pollutants, as well as contributing to a reduction in the CO_2 output. This is examined in detail in the following section.

6.2 Coppice Wood as an Energy Source

S. Nakagawa

Generally, raking up fallen leaves, thinning sprouts from tree stumps, and other activities, such as thinning stands of wood, are indispensable to conserving satoyama landscapes. These activities produce a considerable quantity of fallen leaves, branches, and other biomass resources. Coppices, ridges between rice fields, banks of fields, roof-thatch cropping fields, etc., all produce biomass in addition to that produced in satoyama landscape fields and used originally as the principal resource. Wood has long been used for fuel and tools, and grass was used as compost, livestock feed, and straw bedding in many cases.

The use of biomass resources produced in satoyama landscapes decreased sharply with the collapse of the value of the products. This occurred as a result of the fuel revolution of the 1960s and the rapid decline in primary industries brought about by high economic growth. Although the use of these biomass resources as energy sources will be described in this section, the products of the past coppices will be described briefly first.

6.2.1 Resources in Coppice Woodlands of Old

The vegetation of coppice woodlands in satoyama landscapes changes according to the area. In the Kanto region, most trees are species of *Quercus* (Fagaceae), mainly composed of *Q. serrata* and *Q. acutissima*. In the Tohoku region, *Q. mongolica* var. *grosseserrata* is seen, in part, and, in the Hokkaido region, although the advent of satoyama landscapes is recent, *Q. serrata* can be seen in some areas and *Q. dentate* is seen instead of *Q. mongolica* var. *grosseserrata* in areas such as Tokachi. In the Chubu region, *Q. serrata* is not dominant but *Q. variabilis* is instead. Along the seashore in the warmer districts of the Chubu region, *Q. phillyraeoides*, the evergreen type of *Quercus*, is characteristic of the landscape. In the west of Tokai, Shikoku, and Kyushu, evergreen members of *Quercus*, such as *Q. glauca*, have been used as firewood and charcoal wood. Such a coppice starts with stands that have fast initial growth from seedlings, such as *Mallotus japonicus* and *Cornus controversa*, but is not succeeded by *Q.*

Fig. 6.1. Mature coppice Komorebi-no-Mori in Kanagawa Prefecture

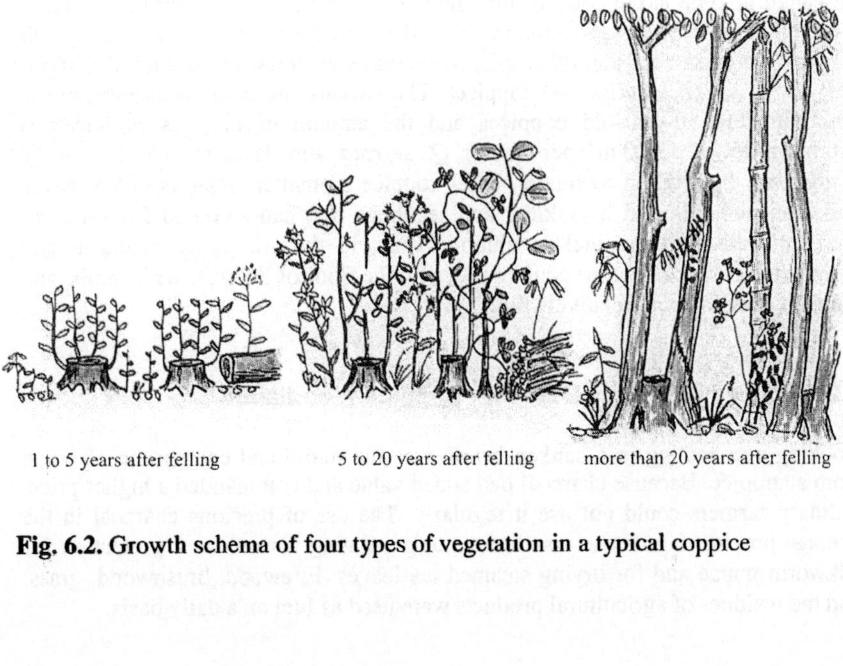

Fig. 6.2. Growth schema of four types of vegetation in a typical coppice

serrata–Q. acutissima (typical coppice) woodlands in many cases. The artificial work of selecting or planting these species and protecting them is necessary for making *Q. serrata– Q. acutissima* woodlands.

Considering Santome Shinden in Miyoshi Town, Saitama Prefecture, and the uplands in the Sagamihara City, Kanagawa Prefecture, where the land was newly reclaimed in the 17th–18th centuries (Figure 6.1), it is known that *Q. serrata* and *Q. acutissima* were planted and coppices were cultivated. As a result of continuous management over 300 years, a variety of patches with different combinations of satoyama landscape plants grew one after another in diverse reproduction stages. A coppice is a complex of plant communities. The *Miscanthus sinensis* grasslands includes *M. sinensis*, *Solidago virga-aurea*, *Aster scaber*, *Angelica decursiva*, *Adenophora triphylla*, *Gentiana sacbra*, and *Cirsium japonicum*. The plant community at the site of coppicing includes *M. japonicus* and *Rhus javanica* in addition to woodland plant species such as *Viburnum dilatatum* and *Cymbidium goeringii* (Figure 6.2). It is thought that *M. sinensis* grasslands, existing prior to coppice development, were made a part of a coppice, and are intact and surviving to this day.

The management of coppice woodlands, including activities such as raking up fallen leaves on the floor of the wood, thinning seedlings that have sprouted from a stump, and felling, which was carried out every 12–20 years, allowed light to penetrate to the woodland floor and allowed the growth of a diversity of sun-loving plants. The inside of these woods was noticeably lighter than woods of *Cryptomeria japonica* or *Chamaecyparis obtusa*.

Next, the production of biomass from coppice woodlands will be estimated. The amount of biomass resource from satoyama woods in Kanagawa Prefecture measured in 1963 in the midst of the fuel revolution was 49.4 m^3 per ha (taken as the average of five sites and assumed to be the same for the rest) in a coppice with 20-year-old *Q. serrata* and other miscellaneous small trees, and a total of 66.6 m^3 per ha in a *Q. acutissima* coppice. The woods have now grown to be approximately 50-year-old coppices and the amount of biomass produced is estimated to be 122.0 m^3 per ha for *Q. serrata* and 153.8 m^3 per ha for *Q. acutissima*. In addition to timber, other coppice biomatter, such as fallen leaves and branches collected by raking, and seedlings that had sprouted from stumps were utilized until the fuel revolution. Thus, in the constantly managed and maintained coppice, grasses could be seen on the floor of the tree, with shrubs and bamboo grasses being relatively rare.

6.2.2 History of Energy Use from Coppice Woodlands

Cooking and heating in a sunken hearth was the traditional energy use of wood from a coppice. Because charcoal had added value and commanded a higher price, ordinary farmers could not use it regularly. The use of precious charcoal in the average pre-war farmhouse was limited to generating warmth as a protection for silkworm pupae and for drying steamed tea leaves. Firewood, brushwood, grass, and the residues of agricultural products were used as fuel on a daily basis.

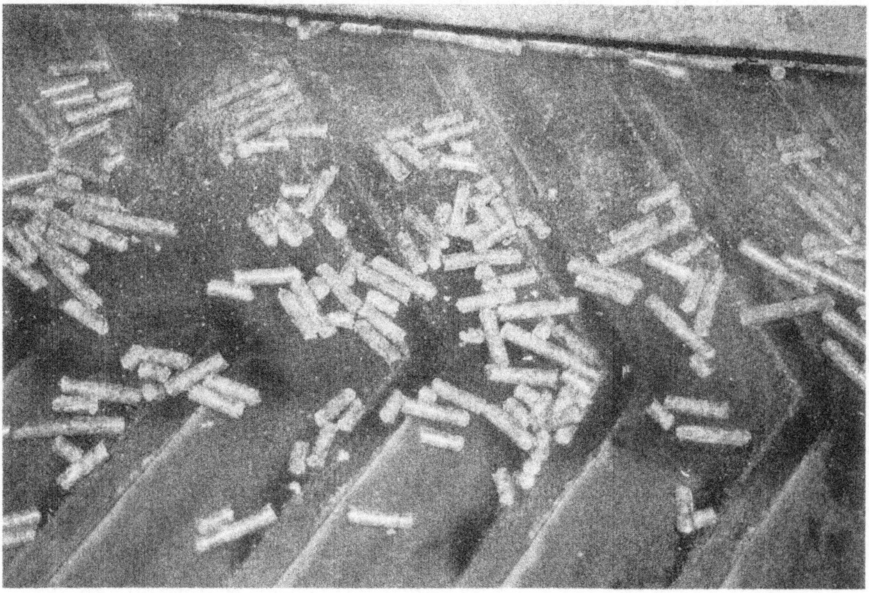

Fig. 6.3. Wooden pellets for fuel in Iwate Prefecture

During the 1940s, the quantity of production of charcoal was the largest. In particular, 1940 saw the production of 2.7 x 10^6 t charcoal. Although the manufacture of charcoal decreased briefly in 1945, the consumption of charcoal continued to increase again until around 1950, when its use as a heat source for heating and cooking decreased abruptly. At present, the quantity of charcoal produced is approximately 28 000 t, which is 1.1% of the peak production. Ironically, the uses for charcoal have increased in water treatment and industrial use, and even now coconut husk charcoal is imported from South-east Asia. In recent years, bamboo charcoal production using a simple kiln, such as an oil drum, has become popular in various places to regenerate bamboo groves that were abandoned. There have been cases in which citizens undertook charcoal manufacture using an oil drum and then used the resulting charcoal as a teaching aid in environmental studies, such as water purification projects.

In 1973, the first oil crisis sounded a warning to the ever-expanding oil consumption. Such was the severity of the situation that the heavy oil for industry and other uses, as well as oil for home use, was difficult to obtain and the use of wood as an alternative was considered. An important development was the system that used wood pellets as fuel. Pellet manufacture was undertaken in the prefectures of Iwate, Yamaguchi, Shimane, and Tokushima, among others. In Shizuoka, heating by wood pellets was introduced in greenhouse horticulture. In Iwate and Tochigi, wood pellets were used a great deal as an industrial energy source, as well as for heating hot spring baths and swimming pools. Pellet manufacture as wood energy and the introduction of a local heating system were seriously considered under a regional plan. However, timber started to lose its competitiveness as the price of oil fell. Despite this, however, pellet manufacture still occurs in Iwate (Kuzumaki Forestry Co.) and in Tokushima (Tsutsui Forestry Co.; Figure 6.3).

6.2.3 Wood as a Renewable Energy Source

At present, domestic consumption of primary energy, such as that used for power generation, is the equivalent of 5.4×10^8 kL crude oil (Japanese Agency of Natural Resources and Energy 1999). The largest proportion of energy is generated from oil (52.4%), the second from coal (16.4%), the third from atomic power (12.3%), the fourth from natural gas (12.3%), and the fifth from hydroelectric power (3.9%). The total sum of low-impact energy sources, including wind power, solar power, and biomass energy, which contributed to 7.5% of total energy consumption in 1955, came to only 1.3% in 1996. However, the absolute quantity of low-impact energy in terms of crude oil increased 2.6-fold from 2.7×10^6 kL in 1965 to 7.1×10^6 kL in 1998. Recently, with the value of biomass energy being recognized, other spin-offs of biomass use have been examined by citizen groups or the private research teams. The Japanese Agency of Natural Resources and Energy targeted 22.8×10^6 kL (3.7% of the total energy consumption) in 2010, which is threefold larger than the absolute quantity of low-impact energy consumption in 1998 (Figure 6.4).

Wood production in the world is 3.38×10^9 m^3 (FAOSTAT 1997) and 1.85×10^9 m^3 (55%) of this is used for fuel (Figure 6.5; Table 6.8). The annual consumption of firewood and charcoal wood per capita is 292 m^3. The Japanese consumption of firewood per capita is 2 m^3 and is among the lowest in the world. On the other hand, the Swedish consumption of firewood and charcoal has increased rapidly to 678 m^3, ranking third after Brazil (679 m^3) and Indonesia (761 m^3) (Table 6.7). In these cases, wood has been used for regional heating systems. In central and northern Europe, the price difference between timber and petroleum is always kept in check by the consumers of these resources, the fuel for home heaters being shifted according to the price of the fuel (Kempten,

Table 6.6 Primary energy in Japan (1000 kL)

	Oil	Coal	Atomic power	Natural gas	Hydroelectric power	New energy	Geothermal power	Total
Mega L	285 267	89 278	74 777	66 995	21 447	6112	1031	544 906
%	52.4	16.4	13.7	12.3	3.9	1.1	0.2	100.0

Source: Agency of Natural Resources and Energy, Japan (1999)

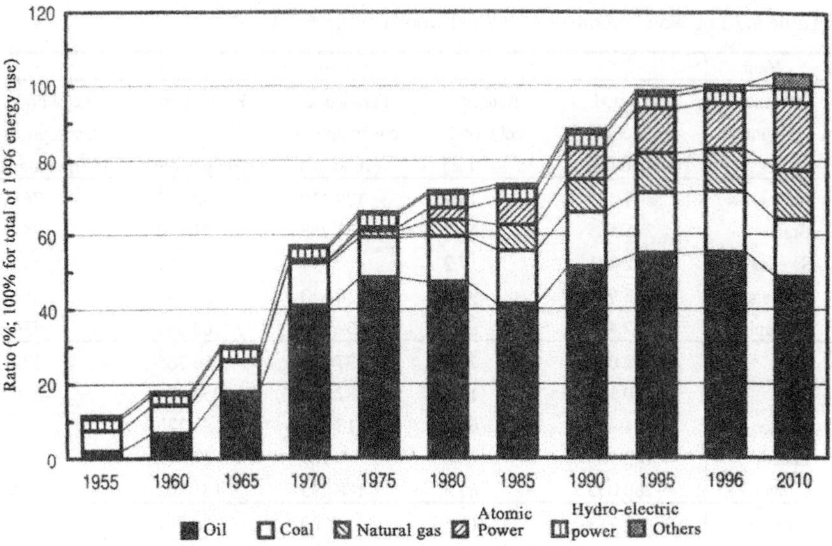

Fig. 6.4. Changes in primary energy use in Japan (1955–2010). (Source: Agency of Natural Resources and Energy, Japan 1999)

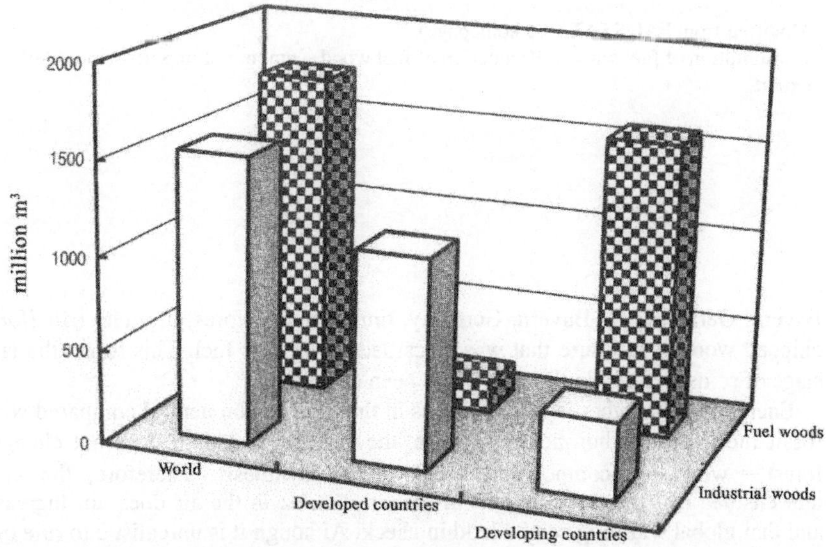

Fig. 6.5. World production of timber (FAOSTAT 1997 homepage)

Table 6.7 Log wood production and fuel wood consumption

Year	1990				
Country	Log wood production (1000 m³)	Rate of fuel wood (%)	Fuel wood consumption (1000 m³)	Population (1000 persons)	Fuel wood consumption (m³/person)
Indonesia	177 485	78.4	139 119	182 812	761
Brazil	205 565	63.9	131 288	147 940	887
Sweden	52 871	7.2	3793	8559	443
Thailand	33 906	90.9	30 813	55 595	554
Philippines	37 463	86.8	32 535	60 687	536
India	262 032	70.7	237 632	850 785	279
USA	510 100	16.3	82 900	254 076	326
Mexico	21 419	64.6	13 843	83 226	166
Canada	162 127	3.8	6169	27 791	222
China	280 033	67.3	188 495	1 161 382	162
Australia	19 913	13.6	2700	16 884	160
Italy	7972	45.6	4054	57 023	71
France	44 713	21.9	9762	56 718	172
Germany	84 707	5.2	4382	79 365	55
Japan	29 403	0.4	103	123 537	1
World	3 319 827	48.9	1 623 506	5 266 445	308

Modified from FAOSTAT (see homepage).
Consumption of fuel wood = Production of fuel wood + amount of import – amount of export.

Bayern, Germany). In Bayern, Germany, timber from a forest or a city (*Alt Holz*; chipped wooden furniture that was discarded) is used as fuel. This forms the last stage of re-use, after all other uses have been exhausted.

Energy use of timber is advantageous in that it is carbon neutral compared with fossil fuels. In a carbon-neutral system, the amount of carbon does not change: forest → wood → decomposition → CO_2 → photosynthesis. Therefore, the system ensures that the total amount of carbon dioxide in the air does not increase and that global warming can be held in check. Although it is unrealistic to rule out the use of fossil fuels for gasoline and plastics in future, the use of timber as an energy source can lessen in the concentration of carbon dioxide in the atmosphere and contribute to the prevention of global warming.

1999				
Log wood production (1000 m³)	Rate of fuel wood (%)	Fuel wood consumption (1000 m³)	Population (1000 persons)	Fuel wood consumption (m³/person)
190 601	83.5	159 243	209 255	761
197 897	57.6	114 052	167 988	679
58 700	10.1	6 028	8892	678
36 631	92.1	33 738	60 856	554
43 399	91.8	39 855	74 454	535
302 794	92.1	278 766	998 056	279
500 745	14.3	71 716	276 218	260
24 122	67.1	16 185	97 365	166
185 659	2.6	4 839	30 857	157
291 330	65.6	190 963	1 274 107	150
22 938	11.8	2700	18 701	144
11 138	62.2	7292	57 343	127
36 380	7.8	2531	58 886	43
37 634	6.8	2559	82 178	31
19 031	1.4	264	126 505	2
3 275 082	53.4	1 747 354	5 978 396	292

Energy use of coppice wood is sustainable and is characterized in Japan by the way in which it forms a contribution to increasing the quality of life, while, at the same time, taking into consideration the natural environment. Satoyama landscape coppices have the characteristic of regeneration. This occurs on tree stumps after felling and does not require any input. Willows were used for such biomass resources in an empirical experiment in Weobley School (Hereford and Worchester County Council), UK; the same coppice regeneration allows this in these coppices, as it does for the poplars of Italy.

The typical afforestation tree species are *Cryptomeria japonica* and *Charmaecyparis obtusa* in Japan. These have the advantage of a higher productivity, from 2.5- to 3.5-fold larger than *Q. serrata* and *Castanea crenata*

Table 6.8 World production of timber (1000 m^3)

	Year	Log wood Total	Log wood Industrial	Log wood Fuel	Sawn wood	Board	Pulp wood	Other industrial	Charcoal (1000 t)
World	1990	3 319 827 (100)	1 696 670 (51.1)	1 623 157 (48.9)	505 278	48 157	423 224	227 169	19 818
	1995	3 212 762 (100)	1 509 072 (47.0)	1 703 690 (53.0)	426 216	55 312	428 235	148 104	21 154
	1999	3 275 082 (100)	1 525 956 (46.6)	1 749 127 (53.4)	431 718	48 084	448 996	156 463	22 567
Developed countries	1990	1 533 054 (100)	1 299 678 (84.8)	233 376 (15.2)	394 884	32 515	362 674	137 861	971
	1995	1 267 790 (100)	1 071 887 (84.5)	195 904 (15.5)	315 384	28 377	354 767	50 694	1273
	1999	1 282 063 (100)	1 117 395 (87.2)	164 668 (12.8)	337 668	27 074	386 289	56 414	1730
Developing countries	1990	1 786 773 (100)	396 992 (22.2)	1 389 781 (77.8)	110 394	15 642	60 550	89 308	18 846
	1995	1 944 972 (100)	437 185 (22.5)	1 507 787 (77.5)	110 833	26 935	73 468	97 410	19 881
	1999	1 993 019 (100)	408 560 (20.5)	1 584 459 (79.5)	94 050	21 010	62 707	100 049	20 837
Japan	1990	29 403 (100)	29 300 (99.6)	103 (0.4)	29 781	6415	10 306	617	37
	1995	23 397 (100)	22 897 (97.9)	500 (2.1)	24 493	4421	5971	446	37
	1999	19 031 (100)	18 767 (98.6)	264 (1.4)	17 952	3261	5024	341	37

Percentages are given in parentheses.
Modified from FAOSTAT (see homepage).

(Table 6.9; Inamura and Nakagawa 2001). However, direct regeneration from seedlings and coppice regeneration cannot be expected for these plantation forests until a number of years have passed. Therefore, the regeneration cost and the use of the biomass of planted woods and secondary woods have to be considered.

Table 6.9 Storage of wood in 40-year-old woodlands

Type of forest	Upper quarter	Middle quarter	Lower quarter	Location
Cryptomeria japonica	419.5	328.5	246.5	Amagi district (Forestry Agency 1965)
Cryptomeria japonica	534.5	394.0	272.6	Abukuma district (Forestry Agency 1955)
Chamaecyparis obtusa	251.5	189.2	127.6	Fuji-Hakone district (Forestry Agency 1954)
Chamaecyparis obtusa	297.0	219.4	158.6	Kanto district (FFPRI 1961)
Quercus serrata	-	101.0	-	Kanagawa Prefecture (Nakagawa 1999)
Quercus acutissima	-	123.4	-	

FFPRI, Forestry and Forest Products Research Institute, Japan.

6.2.4 Coppice Wood as an Energy Source

The caloric value of organic matter is based on the quantity of carbon, hydrogen, and oxygen that it contains. The caloric value of wood is very low compared with petroleum and coal because wood is made up of 65%–75% of carbohydrates, such as cellulose and hemi-cellulose, which are rich in oxygen. The caloric values are 4820–5170 kcal per kg (average 4960 kcal per kg) for dehydrated coniferous trees of Hokkaido (12 species) and 4730 kcal per kg on average for non-coniferous trees (59 species; Abe homepage). Generally, coppice wood contains water (80%–85% for the raw material, 17% for naturally well-dried wood, and 10% for artificially dried material), which significantly reduces the caloric value. The caloric values are 1900–2000 kcal per kg at 50% water and 3800–4100 kcal per kg at 10%–12% water. Therefore, when using coppice wood as fuel, drying is an important stage. Furthermore, consideration must be made of the form that the wood takes to ease transportation, storage, conveyance to a combustion equipment, and ignition nature.

One of the shaped wood materials is chips, which are made by breaking the wood into pieces of nearly fixed size. Chip factories for pulp manufacturing materials can be found in many places in Japan, with the processing systems up and running. Also, to process pruned branches into compost, shredders and chippers have found their way into the gardening industry. Because the wood chips are larger in size and irregular in shape compared with wood pellets (described below), the conveyer that moves chips from storage to the combustion equipment has to be larger in size, with a medium to large combustion equipment being used.

Wood can also be processed as pellets. The main feature of pellets is that their shape is uniform, so that air-conveyers and screw belts can be used. Thus, the use of pellets allows a small conveyance system and a small stove to be used (e.g., for home to a medium-sized boiler system). However, compared with wood chips, wood pellets cost more and processing them is a more complex operation.

The use of pellets has been introduced in Kuzumaki Town, Iwate Prefcture. This town is located to the north of Morioka City in Iwate Prefecture, an area where woods of non-coniferous trees prevail and where the production of charcoal has taken place for centuries. At present, Kuzumaki Forestry Co. Ltd., which produces wood chips for pulp manufacture, is the only producer of pellets. The pellet-production system was introduced in 1980 in order to utilize tree bark produced in chip production. Before the introduction of the system, the bark was piled up out in the open for the production of bark compost or disposed of through incineration.

Water content in the bark and a binding agent are checked prior to the production process. At present, there are a variety of pellet products, with some others undergoing trials (Figure 6.3). The main uses for these pellets are in hot water supply and in the heating of water of swimming pools (i.e., the heating of public facilities and the heating of water at hot-spring resorts).

SumidaTown in Iwate Prefecture, set aside US$80 000 (ca 10 million yen) to promote a project to introduce alternative energy, and went so far as to change the boiler burner in the main town hall and to install a pellet storage tank to allow wood biomass pellets to be used. Furthermore, the local government is planning to introduce pellet boilers in nurseries and is targeting the construction of an independent factory of pellets in the future. Iwate Prefecture has further apportioned US$112 000 (14 million yen) of the 2000 budget for a project to promote the use of wood biomass resources.

6.2.5 Problems and Solutions in Wood Energy Use

As mentioned above, the caloric value of pellets made from wood is approximately half that of petroleum or heavy oil. The price of pellets is approximately US$0.20 (25 yen) per kg and is roughly similar throughout Japan. The price is approximately the same as petroleum products in terms of caloric value, so pellets can compete with petroleum products in terms of price.

In Japan, 15.8 x 10^6 t wood chips is consumed annually, 87% of which is imported (Japan Ministry of Agriculture, Forestry, and Fisheries statistics homepage). Although the price of imported chips varies depending on the tree

Table 6.10 Price of imported and domestic wood chips in Japan

Country	Import		
	1000 t	US$1000	JPY/t
USA	4111	649 669	19 754
Australia	3961	523 891	16 533
Chili	1352	206 807	19 120
South Africa	1302	202 463	19 438
China	680	91 706	16 858
Brazil	542	82 933	19 127
New Zealand	370	53 689	18 138
Canada	363	56 293	19 385
Thailand	174	23 630	16 976
Fiji	151	20 768	17 192
Vietnam	141	21 124	18 727
Indonesia	131	15 802	15 078
Ecuador	118	18 630	19 735
Malaysia	109	11 450	13 131
Argentine	107	17 418	20 348
Russia	61	7791	15 965
Papua New Guinea	50	8612	21 530
Costa Rica	7	1040	18 571
Total	13 730	2 013 716	18 333
Japan coniferous*[1]			16 850
Japan non-coniferous*[1]			22 400

From FAOSTAT (see homepage): calculated as US$1 = Japanese yen (JPY) 125.
*[1]Source: Ministry of Agriculture, Forestry and Fishery, Japanese Government (http://www.hakusyo.maff.go.jp/books_b/WR01H100/html/SB1.4.1.htm)

species and conditions, the average price is US$140 (17 600 yen) per t. The price of domestic chips is US$135 (16 850 yen) per t for coniferous trees and US$179 (22 400 yen) per t for non-coniferous trees (Table 6.10). Furthermore, in Miyazaki Prefecture, wood chips are shipped at US$102 (12 800 yen) per t for coniferous trees and US$169 (21 100 yen) per t for non-coniferous (The Agriculture, Forestry, and Fishery Information Center at Miyazaki Statistics Office of Kyushu Agricultural Administration Office homepage). The supply system of wood chips has already been established and it is thought that there will be no problem in

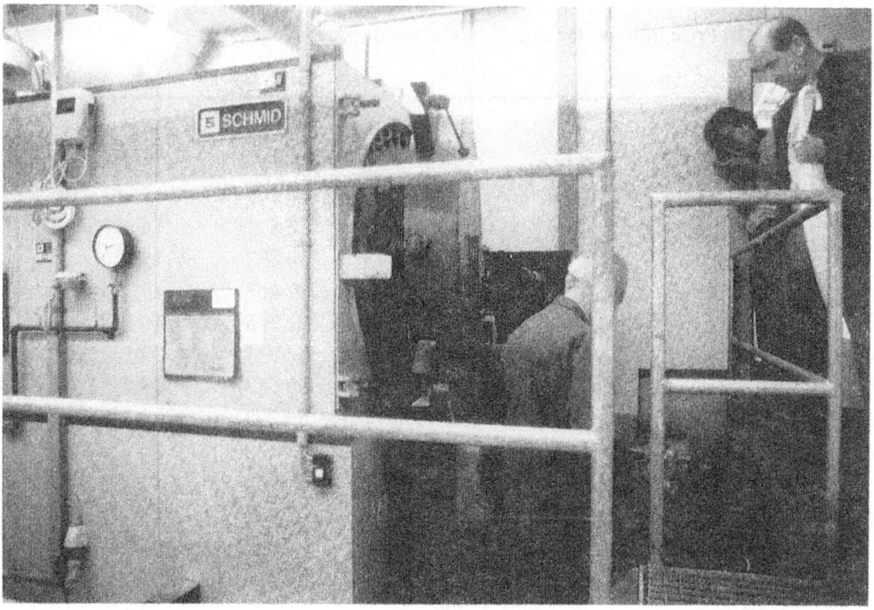

Fig. 6.6. Boiler using wooden chips in Cham, Zug, Switzerland

starting to use it. The biggest problem with wood energy is the disposal of ash, which accounts for 3% of the initial chip weight burned. Petroleum products and natural gas do not have the same problem, so this is expected to be a significant barrier to the future development of wood chip use. Nonetheless, it should be noted that wood ash contains potassium and can be utilized as soil fertilizer. In fact, in Kuzumaki Town, Iwate Prefecture, the ash is used as a fertilizer in the local vineyards. This also occurs at Hamm Town in Zug State, Switzerland, at the Hamm Agricultural Vocational School, where ash is also used as a fertilizer (Figure. 6.6). Further novel uses of ash could be considered in the future.

Although there are currently few systems that utilize wood for energy in Japan, natural gas and oil are used for heating of public facilities in cold districts, such as Hokkaido and Iwate Prefecture and local area heating at Kita-Hiroshima City in the suburbs of Sapporo are applied. Moreover, central heating systems are common in large-sized buildings in the larger cities, such as Yokohama and Tokyo.

It is currently unrealistic to use a wood-burning system in buildings in large cities because of the sheer bulk of the fuel and difficulties with transportation. However, at public facilities in rural cities and in the new towns in the suburbs of large cities, the system might be appropriate. The development of new systems that utilize wood-burning technology would be beneficial to forest management, employment opportunities in forestry, and the economic life in rural outlying areas.

6.2.6 The Future of Energy Use of Coppice Wood

This section will describe the technologies available for using coppice wood as fuel in the near future. First, these materials can be converted into thermal energy, electrical energy, or both (cogeneration) by gasification and, second, they can be converted into methanol using the biological method of microbial fermentation or through a chemical method that uses a catalyst and using the resultant methanol as fuel. The manufacture of electricity through the use of a fuel cell that consumes the methanol has also been considered. Cogeneration has already been shown to be feasible in northern Europe, particularly in Sweden.

The Forest Biomass Energy Research Association, under the Association for Promoting Advanced Science and Technology of Forest Resource Utilization (APAST) described in the previous section, is developing new technologies that efficiently utilize the energy generated from the direct combustion of residual materials from forestry, such as thinned trees, and residual materials generated in lumber mills. The technologies used include gasification by dry distillation and thermal decomposition, small wood gas-generation furnaces, wood residue combustion technology, small steam-boiler turbines, small-scale power generation systems using a gas engine, and small wood gasification cogeneration systems.

As mentioned earlier, the local economy of rural areas can be reinvigorated through the use of a biomass consuming system that includes factories that produce laminated lumber and boards using energy provided from low-quality raw materials, combined with the sale of surplus electricity, and the communal heating of a district. The system, at the same time as being useful in preventing global warming, will maintain the character of satoyama landscapes and reinvigorate the local economy by adjusting the new technology based on the scale and the amount of resources in the satoyama landscapes.

In large-scale satoyama landscapes, as illustrated in the preceding section, 1525 t usable coppice wood can be supplied annually. Supposing that 1000 t from the resource is used for power generation, then the amount of power generated would be 1.8×10^6 kWh per year. This assumes a unit caloric value of coppice wood at 4500 kcal per kg, so 4.5×10^9 kcal is obtained from the 1000 t of resources, the conversion efficiency to electricity is 33%, and 1 kWh is 2450 kcal. Because the average monthly electricity consumption is 300 kWh for Japanese houses, 1.8×10^6 kWh per year of electricity can supply 500 households. Furthermore, if cogeneration that collects and uses heat generated in power generation is used, the collectable heat turns out to be equivalent to a caloric value of 180 kL heavy oil. When comparing this with generated electric power, a reduction of 450 kL heavy oil can be achieved, and approximately 1300 t CO_2 emission from the combustion of 350 kL heavy oil can be prevented.

6.3 Nature Study in Satoyama Landscapes

S. Nakagawa

Regenerated coppice woodlands in satoyama landscapes have various special features compared with coppice woodlands that have not been managed. Both the increase in biodiversity and the potential for use as a biomass resource can be important messages to communicate through environmental education.

6.3.1 Learning Through Management

The following is an example of the ways in which local management and coppice woodland conservation have been included in environmental education programs for local citizens.

1. Stage One: clearing away undergrowth (mainly *Pleioblastus chino* grasses), removal of dead trees and fallen trees, and installation of facilities (paths, barriers etc.) in wooded areas.
2. Stage Two: gaining an understanding, through active participation, of how a coppice woodland functions and what maintenance activities must be performed.
3. Stage Three: restoration of an excessively used site, protection of woodalnd

Fig. 6.7. Volunteer activities at Nanasawa Woodland Park in Kanagawa Prefecture

floor species, production of woodland products such as mushrooms, and community involvement in the activity.

It is known from previous cases in abandoned coppices that woodland floor vegetation begins to change within 3 years of the resumption of periodic management. The former species of plants of satoyama landscapes begin to recover and these can be regarded as indicator species in the management system. Ishii et al. (1993) raised six management types in relation to coppice woodland management in the Kansai region. Various species have been proposed as indicator species of management in other cases of coppice woodland management. The Satoyama Landscape Club in Minami-Kawachi County in Osaka Prefecture regarded *Lilium japonicum* as an indicator species. In the Eco-Forest at Toyota City in Aichi Prefecture, *Magnolia stellata*, *Heloniopsis orientalis*, and *L. japonicum* are all counted as indicator species, and in the JC's Forest at Toyota City, *H. orientalis* is regarded as an indicator species. In the Kanto region, *Viola grypoceras* and *Lilium auratum* are regarded as indicator species by the Kizuna Woodland and Nanasawa Woodland Park Coppice Fan Club at Atsugi City in Kanagawa Prefecture. The restoration of satoyama landscapes in areas that contain these indicator species can be used as ideal environmental education to promote the appreciation of local nature.

6.3.2 Traditional Uses of Wood

Various trees, vines, and grasses have been used to make tools and form the handles of baskets and axes. These tools are called *mingu* or *folk tools* in direct translation. The raw materials for mingu are chosen according to the characteristics of each plant species. For example, *Styrax japonicus* and *Meliosma myriantha* with their straight, flat, and smooth trunks, are used as the handles of scythes. *Pourthiaea villosa* var. *laevis* and *Viburnum dilatatum* are used as hammer handles because they are flexible yet hard to break. The bark of *Tilia japonica* and *Acer rufinerve* is shredded and used as string, bags, and fiber for work clothes (Figure. 6.8). However, the thickness of a scythe handle is limited by the ease with which it can be grasped. When using *S. japonicus*, a branchless section of a 5–8-year-old tree that has sprouted from a stump is used. For a hammer handle, a 10-year-old tree is used and, for sledge hammers, a 1-m length of wood is required for the handle. This can be obtained from a well-grown sprout from a tree stump. When obtaining fiber from *T. japonica* or *A. rufinerve*, the bark of a young tree that is 10–15 years old is stripped. When the bark becomes thicker on older trees, it will become brittle and unsuitable. Thus, in order to use a tree for a specific purpose, both the species and the age of the tree are important. In well-managed satoyama landscapes of old, there were many such variously aged coppices and suitable materials for a specific purpose were easily supplied from coppices in the same area.

However, the currently abandoned coppice woodlands, as a whole, are 30–40 years old and are uniform in age and structure. This makes it difficult to manufacture mingu. Moreover, forestry with locally produced materials is seldom

Fig. 6.8. Harvest of the inner bark of *Acer refinerve* as a traditional utilization of the mountain satoyama landscape

undertaken today. In order to maintain the utilization culture of satoyama environments through the manufacture of mingu, coppice woodlands must be well managed.

6.3.3 Field Activities That Use Biological Resources

The most effective utilization of satoyama landscape resources is that of energy use. In environmental education, cooking using fire is the most intelligible and interesting activity through which to introduce satoyama landscape conservation. It is especially effective for the general public, who are not very interested in satoyama landscapes in general.

An example of an activity that is common in many groups is the cooking of *Baumkuchen* over an open fire (Nakagawa 1997). Baumkuchen is prepared by brushing cake batter onto a rod and roasting it over an open fire. When one layer has become cooked, another layer is brushed on, and this process is repeated until the baumkuchen is the desired size. This was first proposed for use in satoyama landscape conservation activities because it has a surprise factor, it involves cooperation, it requires a specific technique, and it can be expected to yield good results (Figure. 6.9).

Fig. 6.9. Cooking of German traditional cakes, called *Baumkuchen*, in the satoyama

1. Novelty: although many Japanese people know baumkuchen, few people have actually made them or known how to. The novelty of the activity is also brought about by the fact that people think, at first, that it is impossible to make baumkuchen in the woods. Also, people generally do not know how to pull baumkuchen off a bamboo skewer. Although the skewer looks too long to pull baumkuchen off, the baumkuchen can be pulled off using a certain technique.
2. Cooperation: because the work extends over a period of 1–2 hours, several people can be involved, cooperating to complete the work. Several teams of people working in pairs are needed when making baumkuchen: supplying firewood to keep the fire at an ideal temperature and condition, layering the cake mix, baking the baumkuchen by passing them over a fire, and waiting for the change.
3. Technique: participants inevitably compete with others in finishing the baumkuchen because the quality of the results is so clear. The baumkuchen can be over-baked or under-baked depending on the skill of the baker, whereas a fast rotation causes unevenness to develop on the surface and a rotation that is too slow causes the cake mix to fall into the fire. Problems can occur one after another, and there is the fun of learning a technique in a limited period of time.
4. Results can be easily checked at the end of the work. The importance of cooperation can be shown by putting a bamboo skewer out carefully over the

fire, using the hands of participants, so that the baumkuchen does not break. There is always a great deal of satisfaction when, eventually, the hard work is rewarded by cutting the baumkuchen and the rings can be seen clearly.

6.3.4 Use of Firewood in Field Activities

For fuel, newly made firewood is not necessarily required, but small pieces of wood, stands of decayed trees, and dead branches of standing trees are enough. Although the interior of the coppice woodland does not become tidy and bright with one activity, it does after such an activity is repeated several times. By repeating a program in the same place, participants gain an appreciation of how field cooking leads to forest management. Moreover, they can understand that a familiar and usable fuel exists in the coppice woodlands.

Environmental problems can originate from the complex relationship between nature and human beings. Methods at various levels can be used to support this complex web of interaction. Results from solving problems one after another can be like the steady undoing of a knot of tangled thread. Although making baumkuchen is a minor activity in the woods, baking baumkuchen and using firewood as a fuel in the woods can give an insight into woodlot management.

Needless to say, there are two approaches to satoyama landscape conservation for the general public: one that peers deeply into the nub of a problem and its

Fig. 6.10. Firewood preparation, a traditional satoyama activity

causes; and another that allows extensive, but fun, participation in an activity. By integrating these two methods, a citizen's movement becomes a very effective tool for conservation.

Furthermore, an environmental education program that uses wood as fuel in cooking leads to the viewpoint that proper management of woods connects with the lifestyle of the general public who has ignored the management of satoyama landscape coppice woodlands (Figure. 6.10). In addition, understanding can be deepened by knowing that, in Japan, 99% of produced wood is consumed industrially, whereas only 1% of wood is used as fuel. This contrasts greatly with the rest of the world, where 55% of wood is used as fuel, and even in other industrialized countries, where 20% wood is used as fuel.

6.3.5 Prospects and Possibilities for the Future

Coppice woodlands that have been used historically for many purposes have now been abandoned for 30–40 years or more. Landscape features, such as fields, rice paddies, and banks, which provided particular habitats for whole environmental communities, used to be widespread over the satoyama environment. Because of a shortage of farm workers, cultivation and management has been abandoned in some areas, many of which have been changed to artificially planted coniferous

Fig. 6.11. Firewood preparation as a workshop activity in a satoyama

woods or naturally regenerated secondary woods. In such situations, regeneration of satoyama environments is difficult to attain through the sole action of farmers.

The activities of satoyama landscape conservation that resulted from such a historical background are indispensable to the future, although the scale of the activity so far undertaken is small. To appeal widely to citizens to enable the understanding of such ways of thinking, it is desirable that satoyama environments where biodiversity has recovered and satoyama environments where recreational, environmental, and educational activities can take place will spring up. Furthermore, it is also desirable that the general public considers the satoyama landscape as a part of their daily lives.

Although satoyama landscape conservation activities by the residents of a new town can be seen to be taking shape in various places, a greater and more consistent consumption of wood biomass is important because it will prevent management taking place simply for management's sake (Inamura and Nakagawa 2001). It is hoped that doing so will be achieved through a much deeper understanding of satoyama landscapes. For example, the understanding that securing firewood and charcoal as fuel for an emergency in the event of a disaster is the responsibility of city dweller families and local communities.

Taken from such an accessible viewpoint, conservation of satoyama landscapes can result from tangible, everyday activities that produce a mutually beneficial relationship between satoyama landscapes and its nearby citizens (Figure. 6.11).

7 Long-term Strategy for Satoyama Conservation

7.1 Strategic Management of Satoyama Landscapes

A. Tsunekawa

Japanese satoyama landscapes are currently undergoing profound changes. In order to maintain their traditional form, it is necessary to draw up a comprehensive conservation strategy. In terms of planning theory, the strategic content can be divided into two parts. The first part is the subject of quantitative and qualitative management and conservation of satoyama landscapes. As described in Chapter 4, a decrease in biodiversity and the disappearance of rare species have been caused by a loss of the management of satoyama landscapes. Especially problematic are the satoyama coppice woodlands, where the relationship with humans has been noticeably lost.

The second part is the spatial conservation of satoyama landscapes. As described in Chapter 3, satoyama landscapes as a whole, but especially in city suburbs, face an unavoidable and rapid spatial reduction. To protect satoyama landscapes from such an erosion, every possible means, such as land regulation, land taxation, and park open space projects, is needed to comprehensively protect satoyama landscapes.

The first point, how to proceed with both the quantitative and qualitative management and the conservation of satoyama landscapes, will be considered with an emphasis on coppice woodlands. The second point, the spatial conservation of satoyama landscapes, will be described in the following section.

7.1.1 The General Condition of Woodlands in Japan

7.1.1.1 Changes in Woodland Size and Composition

Several statistical data sources will be used to provide an overview of general woodland conditions in Japan. Changes in the extent of the area and the volume of wood in Japanese woodlands (Japan Forest Agency) have been illustrated in Figure 7.1. In 1995, Japanese forests and woodlands were composed of 53.2% natural forests, 41.4% plantation woodlands, and 5.4% of others. Although the forests and woodland area has not changed for more than 30 years since 1966, area in natural forest has increased, while plantation woodland area has decreased by 30%. On the other hand, the volume of wood in woodlands increased by 85% during this period. In particular, the volume of wood in plantation woodlands increased more than

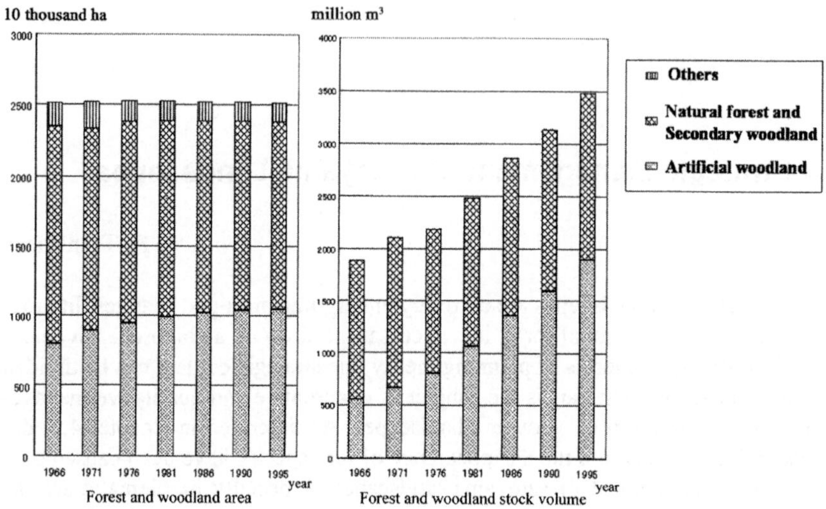

Fig. 7.1. Trends for area and stock volume of Japanese forests and woodlands (Source: Japan Forestry Investigation Committee 2000)

Table 7.1 Transition of regeneration cutting

Year	Area of regeneration cutting: National forest and woodland (ha)	Non-national forest and woodland (ha)	Total (ha)	Area of regeneration cutting/area of woodland (%)	Supply volume of domestic logs (1000 m³)	Area of regeneration cutting/supply volume (ha/1000m³)
1950	213 887	501 919	715 806			
1955	166 952	542 778	709 730		42 794	16.6
1960	141 048	515 181	656 229		49 006	13.4
1965	168 931	402 947	571 878	2.27	50 375	11.4
1970	145 880	339 192	485 072	1.92	46 241	10.5
1975	124 957	234 963	359 920	1.42	34 577	10.4
1980	103 916	219 127	323 043	1.28	34 577	9.3
1985	107 525	183 275	290 800	1.15	33 074	8.8
1990	98 008	145 779	243 787	0.97	29 367	8.3
1995	69 028	89 201	158 229	0.63	22 915	6.9

Sources for area of regeneration cutting and supply volume of domestic logs: Rinya Kosaikai (1982, 1987), Forestry Agency (1997, 2000); area of forest and woodland: Japan Forestry Investigation Committee (2000).

threefold, to 1 900 000 000 m^3 in 1995. Many plantation woodlands developed during the post-war period have arrived at a point where selective cutting is required and will soon reach a point where clear cutting will be required.

However, as shown in Table 7.1, the area of clear-cut woodlands decreased from 720 000 ha in 1950 to 160 000 ha in 1995. Specifically, the ratio of the clear-cut area to total woodland area decreased from 2.27% in 1965 to 0.63% in 1995, and the clear-cut area per wood supply decreased from 16.6 ha per 1000 m^3 in 1955 to 6. ha per 1000 m^3 in 1995. Therefore, because the volume of wood per area increased, the cut woodland area per unit production of wood decreased. Although there is a lack of data regarding the area used for selective cutting in times past compared with the current situation, selective cutting was performed annually on only an average of 210 000 ha in the 5 years from 1993 to 1997 (Japan Forestry Investigation Committee 2000). On the other hand, the coppice woodland area decreased from 1 440 000 ha in 1970 to 990 000 ha in 1980, and to 530 000 ha in 1990.

7.1.1.2 Changes in Energy Consumption and Production of Fertilizer

As has been well documented, satoyama landscapes were long used to provide fuel and fertilizer. The transition in the production of firewood and charcoal is shown in Figure 7.2. Production of both firewood and charcoal peaked in 1955, when the

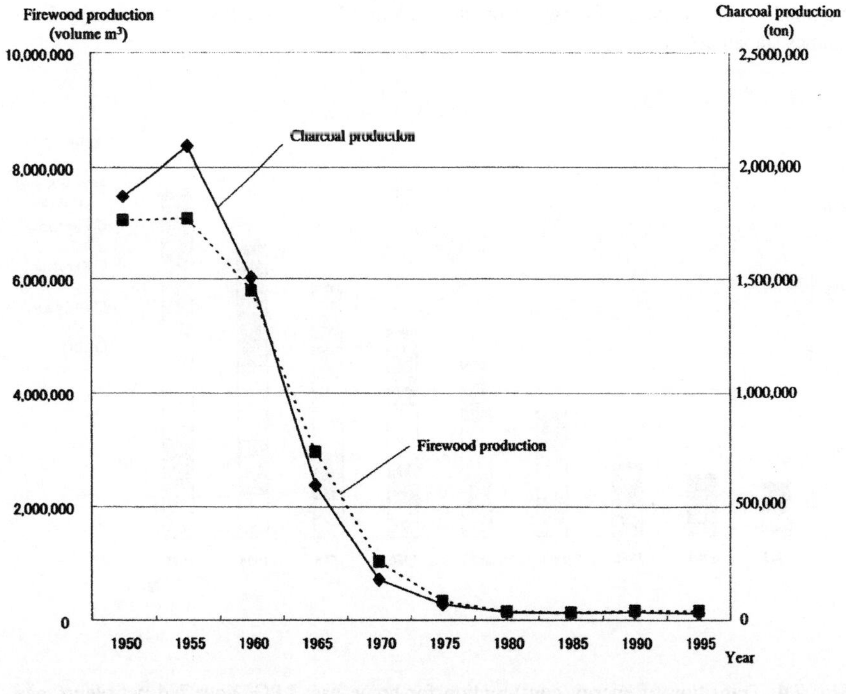

Fig. 7.2. Transition of production of firewood and charcoal (Source: Forestry Statistics Survey, Rinya Kosaikai 1982, 1998, 1992, 1997)

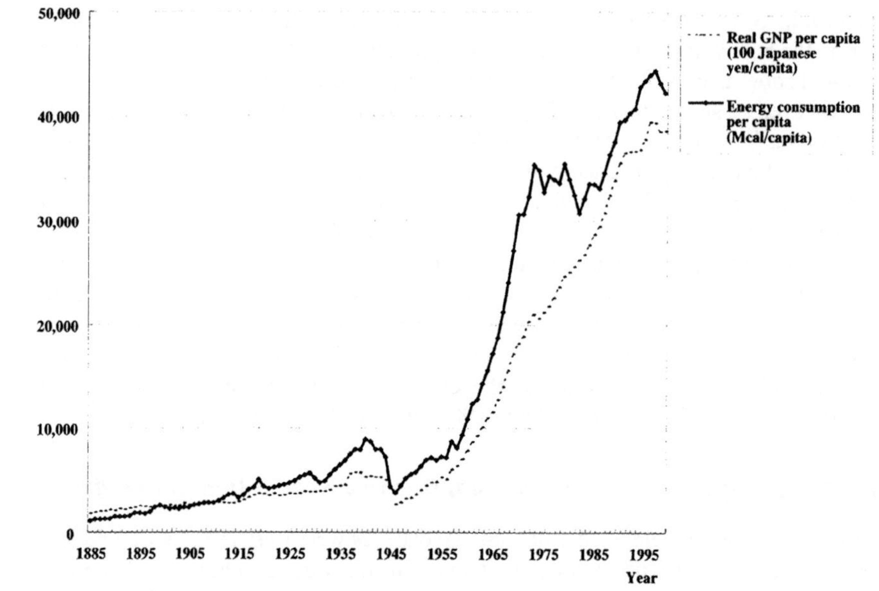

Fig. 7.3. Transition of real GNP per capita and energy consumption per capita. GNP, Gross National Product (Source: Energy Data and Modeling Center, the Institute of Energy Economics, Japan 2001)

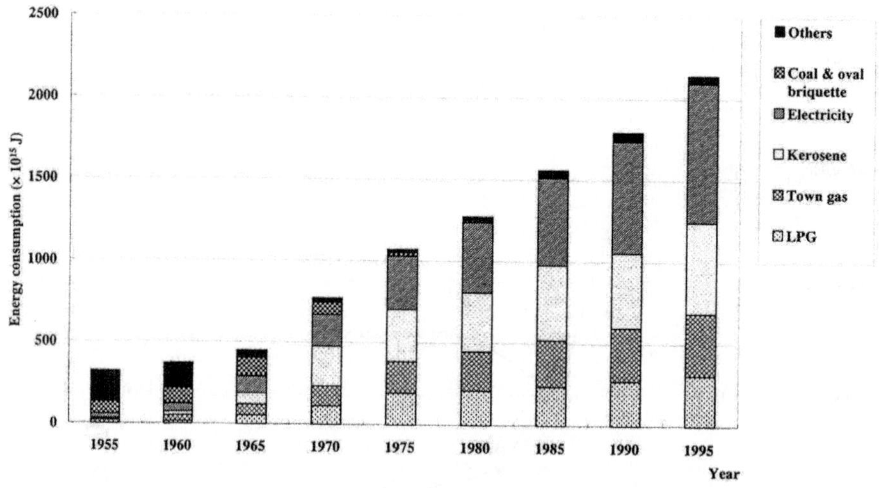

Fig. 7.4. Transition of energy consumption for home use. LPG, liquefied petroleum gas (Source: Comprehensive Energy Statistics, Agency of Natural Resources and Energy 1999)

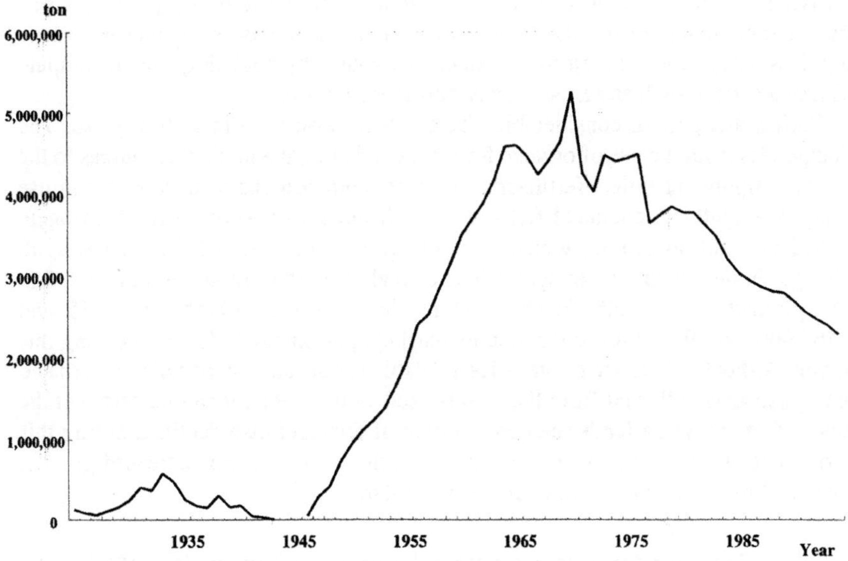

Fig. 7.5. Transition of compound fertilizer production (Source: Pocket Fertilizer Handbook, Ministry of Agriculture, Forestry and Fisheries 1974, 2000)

amount produced was 7 070 000 m^3 and 2 090 000 t, respectively. However, production of firewood and charcoal decreased rapidly to 150 000 m^3 and 3500 t, respectively, in 1980, with the numbers leveling off since then.

The decrease in the production of firewood and charcoal since 1955 is explained through the change in Japan's energy consuming structure as a whole. In Figure 7.3, changes in the real Gross National Product (GNP) per capita and energy consumption per capita are shown. From this figure, it can be seen that there is a strong correlation between the two. Both GNP and energy consumption have shown significant growth since 1955, exactly as the production of firewood and charcoal began to decrease. This period is called both the period of high economic growth and the energy revolution. The amount of energy consumed per capita increased approximately fourfold from 7183 Mcal per person per year in 1955 to 30 546 Mcal per person per year in 1970. The increase in energy consumption slowed after that, with the consumption being 42 161 Mcal per person per year in 1999.

The energy revolution points to a shift from the use of charcoal to petroleum, which was seen globally after the war. The direct factor in the energy revolution is the economic predominance of petroleum over charcoal. Some of the causes of this were the discovery of large oil fields, the introduction of huge tankers, and the development of long-distance, large-diameter pipelines.

The transition of home energy consumption is shown in Figure 7.4. In 1955, the consumption of *others*, namely firewood, charcoal etc., accounted for 58% of the total energy consumption. However, by 1970, this figure had dropped to 4%, while kerosene and electric power increased. Although cheap energy supply was a primary factor in supporting the high growth of the economy, satoyama landscape coppices inevitably lost their role as an energy source by not being able to compete with other fuels, such as kerosene, gas, and electric power.

During this period, considerable changes were also seen in fertilizer use. The change was from the use of organic fertilizers and animal and plant resources to the use of inorganic chemical fertilizers. The most common chemical fertilizer in use today is a synthetic chemical fertilizer, which includes two or more of nitrogen, phosphate, and potassium, which account for 15% or more of the total weight. Changes in the production of synthetic chemical fertilizers are shown in Figure 7.5. The production of synthetic chemical fertilizers was 1 110 000 t in 1955 and 4 460 000 t in 1970, with production increasing approximately fourfold during this period. Although there are no statistics available for the amount of fallen leaves and other materials collected from the woods used as fertilizers, it can be said that the function of satoyama landscapes as a source of fertilizer also declined during this period because the amount of chemically synthesized fertilizer increased greatly, whereas farmland area remained roughly constant.

7.1.1.3 A Turning Point in Satoyama Landscape Wood Use: 1955–1965

As noted previously, the mode of use of Japanese woodlands, especially satoyama landscapes, changed drastically after World War II. Specifically, this occurred from 1955 to 1965, the turning point in satoyama landscape wood use. During this period, home energy consumption per capita increased 1.4-fold and the main energy sources shifted from firewood and charcoal to petroleum, electric power, and gas. Fertilizers shifted from organic fertilizers to chemical fertilizers, especially synthetic chemical fertilizers, and the production of synthetic chemical fertilizers increased 3.3-fold.

Under such conditions, the management of satoyama landscape woods, once performed regularly by villagers, was abandoned. This means that the variety of biota and the previous form of satoyama landscapes are being lost, and can only be conserved through appropriate management.

7.1.1.4 Who Should Manage Satoyama Landscapes?

The satoyama woodland that had traditionally been maintained regularly by generations of local residents through their agriculture and forestry activities, has now been essentially abandoned. In an attempt to conserve various plants and animals that have come to rely on the managed woodlands, a reimposition of suitable management is being increasingly called for. So, who should manage a village woodland, and how should it be done?

7.1.2 Management Approaches for Satoyama Woodlands

7.1.2.1 Management of Satoyama Landscape Woodlands

There are essentially two types of management of satoyama landscape woodlands: (a) management performed every year; and (b) management performed approximately every 15–20 years. The former type of work is mainly weeding, raking up fallen leaves, and other activities, such as composting leaves. Other types of work include felling trees and the selective cutting of seedlings from a stump. These activities serve as a way of dynamically conserving satoyama landscape woodlands as a whole.

7.1.2.2 Felling and Weeding by Forestry Workers

In 1995, the total area of clear cutting and selective cutting was 158 000 ha, whereas the total volume of wood produced was 29 285 000 m^3 (Forestry Agency 2000). The area over which trees were thinned in 1995 was 215 000 ha, with a volume of 41 000 m^3, including 18 000 m^3 of wood consumed (Japan Forestry Investigation Committee 2000). The coppice woodland area in 1990 was 530 000 ha. Therefore, of the 25 250 000 ha total woodland area in 1995, the proportion of the area in which clear cutting took place was 0.6%, the area of selective cutting was 0.9%, and the area of coppicing was 2.1%. However, the extent to which these activities have occurred in satoyama landscapes is unknown. Because of the economics of forestry, most of the managed woodlands are probably timber woodlands composed mainly of conifer stands.

7.1.2.3 Management of Satoyama Woodlands by Volunteers

The Nature Conservation Society of Japan conducted an investigation into actions that establish a rapport with satoyama landscapes that was commissioned by the Japanese Ministry of the Environment. For the investigation, the society distributed questionnaires to approximately 3000 satoyama-related activity groups nationwide and collected questionnaires from 1013 groups and individuals (National Parks and Wildlife Office of the Ministry of the Environment and Nature Conservation Society of Japan 2001). As a result of this investigation, it was found that 373 groups and individuals were performing activities related to the maintenance and management of coppices and grasslands, with the total number of participants being 7631. However, the list of coppice woodland regeneration and conservation groups shows approximately 12 000 active members and 11 000 supporting members nationwide. Moreover, because there are groups that are not listed, it may be presumed that the actual number of citizen volunteers is larger. Although the number of citizen volunteers participating in the management of satoyama landscape woodlands varies in terms of the activity and type of management, it is likely that there may be as many as 20 000 people in Japan who are involved.

As a general rule (Nakagawa 1996), it takes 10 people to manage each hectare of satoyama landscape woodland. Assuming that there are 20 000 citizen volunteers participating in the management of satoyama landscape woodlands, approximately 2000 ha could be expected to be managed by these voluteers. Given that approximately 20% of Japan (approximately 60 000–90 000 km^2) is in satoyama landscape woodlands, the area that could be managed by citizens at present can only be expected to be, at most, 0.03% of all satoyama landscape woodlands (using the conservative value of 60 000 km^2). If even 10% of satoyama landscape woodlands were managed in this way, 600 000 ha would require an army of 6 000 000 volunteers.

7.1.2.4 Management Within a Park

In some cases, satoyama landscapes are being managed, and their woodlands maintained, by commissioned gardening companies. An example of this is the Musashi Hills Woodland Park in Saitama Prefecture. The maintenance and management expenses (for the fiscal year 2000) of woods in the Musashi Hill Woodland Park are shown in Table 7.2. This park has an area of 304 ha, an admission fee of US$3 (390 yen for an adult in June 2000) is charged, and a little under 1 000 000 people enter the park annually. Of 176 ha of woods in total, the areas, according to type, are 153 ha of coppice broad-leaved woodland, 15 ha of *Pinus densiflora* woods, 2 ha of *Cryptomeria japonica* woods, 1 ha of *Chamaecyparis obtusa* woods, and 5 ha of bamboo woods. Of a total of US$428 000 (53 500 000 yen) spent on maintenance and management, US$310 400 (38 800 000 yen) was spent on weeding, US$56 000 (7 000 000 yen) was spent on thinning, US$5600

Table 7.2 Woodland maintenance and management costs at Musashi Hill Woodland Park in 2000

		Cost (US$1000)
Weeding area (m^2)	457 000	310
Thinning (no. trees)	4400	56
Bamboo cutting (no. bamboos)	2000	6
Disposal of dead red pine trees (no. trees disposed)	550	28
Prevention of death of red pines (no. trees disposed)	870	28
Total		428

125 Japanese yen ≈ US$1.
Source: Parks and Recreation Foundation (2001).

(700 000 yen) were used for felling bamboo, US$28 000 (3 500 000 yen) was spent disposing of dead pines, and US$28 000 (3 500 000 yen) was spent preventing the death of *Pinus*. Because the wooded area is 176 ha and an average of US$2400 per ha (300 000 yen per ha) was spent, US$1760 (220 000 yen) was spent on weeding and US$320 (40 000 yen) was spent on thinning annually. If management was commissioned to contractors in order to maintain 10% of the satoyama landscape woodlands in Japan (600 000 ha) as they are in the Musashi Hill Woodland Park, US$144 000 000 (18 000 000 000 yen) would be required, with US$105 600 000 (13 200 000 000 yen) required for weeding alone.

7.1.2.5 Management Feasibility

As mentioned above, the area managed by woodland workers is approximately 25 000 000 ha of the Japanese woodlands, which amounts to 0.6% in clear cutting, 0.9% in coppicing, and 2.1% in weeding. Furthermore, the area managed by citizen volunteers is probably less than 0.01%. Therefore, at present, management by woodland workers accounts for a substantially larger amount, so it is impossible to expect citizen volunteers to do much of the actual work. However, even if all the management by woodland workers, citizen volunteers, and commissioned contractors could be mobilized to take place, the total amounts to very little. The reality is that most Japanese satoyama landscape woodlands are abandoned and are likely to remain that way. Therefore, developing a management strategy with the greatest effect and minimum human and economic expenditure is necessary.

7.1.3 The Potential for Biomass Use

7.1.3.1 Biomass Disposal

In managing satoyama landscape woodlands, how should the biomass produced, namely mowed bamboo grasses and thinned out trees, be used or disposed of? Biomass used to be a precious resource as fuel, compost, feed, and as a raw material. However, as has already been discussed, its value has declined relative to its former uses. Furthermore, a serious problem that has recently come to light is that of dioxins emitted during the burning of waste in the field. Dioxins, polychlorodibenzofuran, polychlorodibenzo-*p*-dioxin, and coplaner polychlorobiphenyl, are generated during the combustion process when garbage or refined metals are incinerated, and through the chlorine bleaching process of paper making. It is said that the dioxins can cause loss of weight, dermatitis, liver disorders, neurosis, and reproduction toxicity in people. In the Law Concerning Special Measures Against Dioxins enacted in January, 2000, with the revised Offensive Odor Control Law, the regulations governing the burning waste in the field were tightened sharply.

How should the biomass be disposed of when burning in the field is prohibited? If the biomass is left, it will turn into an eyesore. Moreover, carrying biomass to a garbage incineration facility will be too labor intensive and costly. According to internal data of the Japanese Forest Agency, the volume of thinned trees was 45 300 m^3 in 1997 and, of this volume, 26 000 m^3 was used; that is, the utilization rate was only 43% (its main use was for lumber; Japan Forestry Investigation Committee 2000). In management performed by citizen volunteers, although thinned trees are used as charcoal and bed logs for the cultivation of mushrooms, most of the bamboo grasses are probably left where they are after mowing.

Although abundant information about biomass as a resource has already been provided in Chapter 5, the following is a discussion of how to consider biomass obtained from satoyama landscape woodlands throughout Japan as a whole. Is there a way of not treating biomass as waste but utilizing it as a resource?

7.1.3.2 Use of Satoyama Resources as Energy

As discussed in Chapter 5, it is assumed that the average accumulation of Japanese woodlands is 120 m^3 per ha, the annual growth rate is 5% (6 m^3 per ha), and the available amount of supply for achieving continuous production is 4 m^3 per ha (2 t per ha when the specific gravity is 0.5). The area of Japanese satoyama landscape woodlands is 6 000 000 ha and 10% of the woods, or 600 000 ha of satoyama landscape woodlands, will produce 1 200 000 t (12×10^9 kg) of firewood. Assuming the unit caloric value of wood resources as 4500 kcal per kg, the total caloric value translates as 12×10^9 kg × 4500 kcal per kg = 5.4×10^{12} kcal. Because the primary energy aggregate supply (for the fiscal year 1999) was 548 649 × 10^{12} kcal (The Energy Data and Modeling Center of the Institute of Energy Economics, Japan 2001), the continuous organic production obtained from satoyama landscape woodlands is equivalent to 0.1% of Japan's primary energy aggregate supply.

This shows that it is unreasonable to expect the diffuse scattering of satoyama landscape woodlands to be a viable source of energy supply. Further, even if the energy can be supplied from satoyama landscape woodlands at a low cost, it cannot be expected that the energy price will fall as a result of excessive supply.

Now, assuming that 2 t per ha of biomass can be obtained from satoyama landscape woodlands annually and that the unit caloric value is 4500 kcal per kg, the annual energy will be approximately 2.0×10^3 kg per ha × 4.5×10^3 kcal per kg = 9.0×10^6 kcal per ha = 9.0 Gcal per ha. The cost of energy production was estimated solely from the cost of cutting, ignoring the cost of conveyance, processing etc. Assuming the cost of cutting as US$1600 (200 000 yen) per ha, this translates as US$0.17 (22 yen) per Mcal and, if the cost of cutting was US$3200 (400 000 yen) per ha, the total cost becomes US$0.35 (44 yen) per Mcal. Moreover, according to the Forest Agency, there is no standard value that reflects the cost of clear cutting, thinning, and weeding. For example, the Forest Agency is subsidizing a thinning project. The amount of the subsidy and consequent actions vary among prefectures; the standard cost cannot be estimated by averaging these subsidies (the same is true for weeding and clear cutting). As a concrete example of the amount of money used, in Nagano Prefecture, when the age of trees is between 16 and 35 years, the subsidies offered per hectare amount to US$1128 (141 000 yen) for thinning and leaving harvested materials, and US$1904 (238 000 yen) for thinning and removing materials. However, if the assessment coefficient (0.9–1.7) is applied according to the political importance, the actual subsidies vary even more.

When using fuel efficiency estimated by average consumption as the base, the required cost for obtaining 1 Mcal of heat is US$0.04 (5 yen) for kerosene, US$0.11 (14 yen) for natural gas and US$0.22 (28 yen) for electricity (Oil Information Center, Japan). Compared with these prices, even if ignoring the cost of conveyance and processing, there is a price differential of 4–9-fold for kerosene, 1.6–3-fold for natural gas, and 0.8–1.6-fold for electricity. These calculations assumed that 2 t per ha of biomass could be obtained with a cutting cost of US$1600–3200 (200 000–400 000 yen) per ha. That is, the cost per ton of biomass is US$0.80–1.60 (100–200 yen). In comparison with the price of the domestic production of wood chips, at US$134.80 (16 850 yen) per t for coniferous tree chips and US$179.20 (22 400 yen) per t for broad-leaved tree chips, this cost is 10-fold too high. This difference is because the former is thinning in satoyama landscape woodlands and the latter is clear cutting in mainly afforested woods. The estimation of the production cost varies greatly depending on which cutting method is assumed to be used, as shown in these cases.

7.1.3.3 Use of Satoyama Resources as Raw Materials

Most Japanese woodlands currently supply raw materials. As discussed earlier, the Japanese woodlands amounted to approximately 3 500 000 000 m^3 and increased by approximately 2% (68 000 000 m^3) every year (Figure 7.1). The amount of wood supply and demand (for the fiscal year 1996) was 114 040 000 m^3; 90 276 000 m^3 was imported and domestic production was 23 764 000 m^3. As for items of demand, lumber (49 758 000 m^3; 44%), pulp chips (43 822 000 m^3; 38%), and plywood (15 726 000 m^3; 14%) constitute the vast majority (Forestry Agency 2000).

Similarly, assuming the annual growth of woodland to be 4 m^3 per ha, 2 400 000 m^3 of wood production can be obtained from 600 000 ha (10%) of satoyama landscape woodlands. This is 2% of Japanese wood supply and demand, equivalent to 10% of domestic wood production.

The price of wood products (in 2000) was US$137.60 (17 200 yen) per m^3 for medium logs of *Cryptomeria japonica* (14–22 cm in diameter and 3.65–4 m in length) and US$61.60 (7700 yen) per m^3 for log pulp of broad-leaved trees. Suppose that a tree with a 15-cm diameter and 10 m in height is felled, and the cost of felling is US$10.40 (1300 yen) per tree (based on the thinning cost in Musashi Hill Woodland Park). The part that can be used as a resource is only a straight, 4-m length of trunk:

$$(0.075\ \text{m})^2 \times 3.14 \times 4\text{m} = 0.07\ \text{m}^3$$
$$0.07\ \text{m}^3 \times \text{US\$}137.60 \text{ per m}^3 = \text{US\$}9.63$$
$$(0.07\ \text{m}^3 \times 17\,200 \text{ yen per m}^3 = 1204 \text{ yen})$$

Therefore, US$10.40 (1300 yen) is required to fell the tree and, in the end, it is worth approximately US$9.60 (1200 yen). For broad-leaved trees, the yield rate is even lower, with only approximately 2.5 m of the 10 m being usable, resulting in a value of only approximately US$2.72 (340 yen).

7.1.3.4 Satoyama Landscapes As a Measure Against Global Warming

As mentioned earlier, only approximately 5% of Japanese satoyama landscape woodlands are managed and forestry workers manage many of these areas. A reasonable target for management might be 10% or approximately 600 000 ha. This would be a challenge, but it is thought that a logical management scheme could be drawn up and it would be possible to use or dispose of the biomass produced. Agricultural and forestry traders with rich experience and practical knowledge would be principally responsible for the management of the satoyama landscape woodlands. One of the schemes to induce these people to manage satoyama landscape woodlands would be to provide them with an economic incentive.

Because the area of woodland managed by citizen volunteers is less than 0.1% of the total area of satoyama landscape woodlands, it cannot be expected that citizen volunteers will be entrusted with the management of all satoyama landscapes. However, because the circle of citizen volunteers participating in the conservation of satoyama landscape woodlands is getting larger, such citizen understanding of the importance of these woods may lead to a renewal of people's understanding of the value of these areas throughout Japan and may lead directly and indirectly to the management and utilization of the woodlands by citizen volunteers. Therefore, positively supporting the activity of satoyama landscape conservation through citizen volunteers is important.

Furthermore, considerable expenditure would be necessary to hire landscape gardeners to manage satoyama landscape woodlands. However, almost 1 000 000 people enter Musashi Hill Woodland Park, even though there is an entrance fee. For satoyama landscapes that are easily accessible from cities and have fine scenery, it might be a good idea to transform the satoyama landscapes into parks and entrust private contractors with their management. However, the management of satoyama landscapes is best based on traditional management methods, which differ from standard landscape gardening practice. Moreover, it is also necessary to consider the biomass that is produced. Biomass can be used as a raw material and for energy. In reality, 40% of thinned trees are used, with most being used as raw materials. However, the methods of biomass use will probably diversify in time, such as making wood chips from scrap or used wood and using them as energy. Biomass is a renewable resource that needs to have its effective use placed in a societal context, as well as it being necessary to assess the use of the biomass in terms of its potential to prevent global warming.

Although the examination here was made in general for satoyama landscapes across Japan, the discussion can be extended to a local case study. Much important research is currently being targeted toward studies of the development of natural and socioeconomic evaluation methods of satoyama landscape woods: comparing management costs with the benefits that can be accrued from the management of such areas; examination of the partnership among villagers, citizens, and administrations; evaluation of the theory of such planning to recommend better management; or cooperation with spatial reservation policies and plans, and coordination with other land use and forestry plans. However, further research is necessary in all these areas.

7.2 Legal Systems for Satoyama Landscape Conservation

Y. Kitamura

7.2.1 Strict Regulations Can Limit Creative Options

The Japanese legal system of land use is based on the idea that development is free to take place as long as there is no concrete regulation system. Moreover, even if there is a regulation system, the guiding principle has been that development is acceptable if the national uniform rule laid down by the central ministries is followed.

There is very little legal support for the conservation of landscapes. Strict regulation systems with environmental conservation in mind, such as the Nature Conservation Areas enshrined by the Nature Conservation Law, cover only a few, relatively small areas. In addition, even if an area receives Nature Park designation, if it is classified as an ordinary area, the regulations are not very strict. Moreover, the legal system examines a development by applying a predetermined national standard. If the standard has been met, the development will be permitted. Therefore, although the legal system is set up to restrict inappropriate activities and developments, it also disallows creative problem solving and limits non-standard development. This is further complicated by the fact that respect for property rights is a basic principle in Japan and this, too, is firmly ingrained to allow a strengthening of standards to a useful degree.

There are many laws related to land use. Each law's regulations pertain to a specific legal objective, with mutual cooperation among them not necessarily being achieved. Although it is necessary to obtain several legal permits to carry out a certain development, development is not really hindered because the requirements of each level of government are not very stringent. An analogy with university graduation can be drawn: it is possible to graduate by gaining credits with a pass rather than a good or an excellent mark. In any case, it is not possible to say even that the resulting development will suit the local environment. Despite the existence of excessive development, a term often used to describe certain kinds of development following the enforcement of the Japanese Resort Law, most excessive developments were not considered illegal developments. They were lawful, but the development that resulted was not environmentally sound.

7.2.2 The Philosophy of Land-Use Regulation

Japan is a country of great diversity, both physically and socially, so uniform standards do not necessarily result in appropriate developments at a local scale. However, because work related to land-use regulation was the office work of a country, municipalities began to regulate development acts, adding standards, and developing discretionary rights for regulation. On the other hand, the Central Government maintains the stance that uniform regulations by law are appropriate for land use because of the security of property rights.

Some municipalities adjusted the development acts to suit the local environment by establishing regulations that precede the legal procedures of the Development Act. These procedures are termed land-use regulations and have been developed in cooperation with appropriate administrative departments. A development project is examined based on an autonomous environmental plan, land-use plan etc., and, where necessary, administrative guidance may be brought to bear to correct a development that is thought to be inappropriate. As a result of the administrative guidance, developers voluntarily follow the guidelines; however, because it is not a strict legal procedure, the developers are under little obligation to follow the guidelines. Development projects will not necessarily be in accord with the autonomous environmental plan, land use plan etc. In some cases, a hearing of citizens' opinions, which is not accounted for in legal procedures, will also occur.

There is a macro- and a microside to land-use regulation. Adjustments made to an individual development plan account for the microside. However, the macroside is accounted for by original measures, which are taken according to opinions received from a range of stakeholders, enacted on a legal basis, but taken when the law is not sufficient to deal with local environmental management.

The laws of a nation tend to be divided vertically because of the relationship among the jurisdictions of various agencies. However, the natural and social functions of the environment exist outside this organization. Municipalities, which can develop comprehensive administration under one leadership, have the ability to correct laws that are not appropriate. Such a method through land-use regulation is also effective for satoyama environment conservation. In order to enrich this scheme in future, it is necessary to consider some of the following points.

7.2.3 Land-Use Regulations During the Decentralization of Governance

The first point is the link with the decentralization of governance. In the past, it has been a principle that a municipality was not able to exercise any authority based on local conditions, because the work of a municipality was actually the *work of a nation*. This system, called the agency delegated function system was abolished during the decentralization reform. Consequently, the conventional functions of the central government were delegated to the offices of the municipalities, and the establishment of local laws was allowed. Although additional regulations were required, it enabled the administrative guidance to be followed more rigorously. Thus, as a result of this decentralization, a municipality was able to tighten up any environmental regulations it saw fit through the enactment of local laws. It also became possible to expand the legally restricted area through a separate policy.

The second point is legislating for a matter that cannot be evaluated quantitatively. For example, it is not appropriate to judge scenery with an all or nothing numerical standard. To process such a case, a mechanism that involves people across a region to argue for the suitability of certain guidelines is required. Because the guidelines will determine the environmental image of an area, citizens must be actively involved in the process.

The third point is how to treat the infringement of a guideline. Until now, development entrepreneurs were only notified. Because the suitability of a development was not checked by a legal permit-based system, it was legally possible to continue a development plan regardless. This situation arose because of a law that caused the link between the legal procedure before the application for a permit and the national-level procedure after the application to be broken. Such restrictions were also cleared by the abolition of the agency delegation function system.

7.2.4 The Land-Use Side of Satoyama Landscape Conservation

What points can be raised in relation to land use and the conservation of satoyama landscapes? These can be organized into three categories. The first is the relation of land use to an adjoining city area. If a city expands in a disorderly way, it will have a big influence on the satoyama landscape, something that will become a problem to control. The second is to develop an understanding of a continuous ecosystem network that even goes so far as to stretch from coastal plains to the mountains. The third is the importance of recognizing land use in terms of conserving farmland and revitalizing villages. Because satoyama landscapes are places where people live, revitalizing a village as a local community is important.

7.2.5 Requirements of a Land-Use Regulation System

What roles are required of a land-use regulation system? This can be summarized by the following points. The first point is an elimination of unregulated areas. As such, in satoyama environments, some of the areas are not included in the city planning zone or are not restricted even when included. These areas are commonly termed blank areas. Because legal regulation cannot be fully undertaken in many cases for these areas, it is necessary to consider the elimination of blank areas at a macrolevel of land-use regulation. For example, for blank areas in a city planning zone, application of the specific land-use restriction zone, which was introduced by the revised Town Planning and Zoning Act in 2000, and application of the quasi-city planning area, also introduced by the Act, can be considered. The specific land-use restriction zone is a mechanism that restricts land use, such as large-scale stores and pachinko parlors, by providing for city planning in blank areas. Quasi-city planning areas are areas that may potentially cause trouble in terms of maintenance, development, and conservation if the current land use is left to continue as it is and the land use in the surrounding area of existing villages is not regulated. These zones can be designated at the municipal level.

Second, land use plans concerned with satoyama environments need to be clarified. When trying to propose a development act in a satoyama environment to suit the environment, guidelines must be defined clearly and the future image of an area designated clearly. Moreover, without appropriate regional characteristics, a guideline cannot be referred to by developers, causing difficulties in the dialogue between stakeholders, as well being difficult to administer in the end. In a planning decision, methods such as citizen participation in a workshop should be utilized. If these region-specific plan contents are drawn up according to a democratic decision procedure, they will be bound by a force that guarantees their legitimacy.

Third, a mechanism should be structured that enables arguments to be heard about an individual development project from the entire local community. A developer has to explain how a development relates to a land use plan when presenting a planned proposal. Residents raise questions and arguments according to the land use plan to be drawn up, and a developer replies to these using a standard procedure. This procedure should be clear and fair.

The fourth point is that of planning to activate a local area. Almost any area over which a plan will be used will contain parts with strong development pressure and also parts that show problems of local promotion, desolation of farmland and woodland, and so on. A land-use regulation system should be formulated that recognizes the differences in these areas through an independent consideration of the future development in an area and what the land uses should be. In a land-use regulation system, the necessity of the establishment of a satoyama community should be recognized independently and actively.

Finally, the system should become highly responsive, where citizen participation is fully secured and the results are reflected in the construction and management of an institution. Recognition of the various functions of satoyama environments and a decision-making authority is not necessarily only a matter for satoyama residents. On the other hand, city residents alone may not decide these either. Although it is natural that satoyama residents' intentions are fully represented, because a major premise is the maintenance of communities and their property, participation of both satoyama residents and citizens living outside the area should be regarded as important in the decision-making process regarding most plans and the evaluation of development projects because, in reality, satoyama society is formed with support from citizens living outside the area.

7.2.6 Satoyama Landscapes and Land-Use Regulation

In order to tackle the above subjects, satoyama functions must be identified, as should the image of what satoyama environments should become. These should be postulated by using planning that includes a concrete direction and character for the conservation efforts. Planning that satisfies these points will be utilized in regulating land use. There is scope for the selection of the spatial extent covered by the

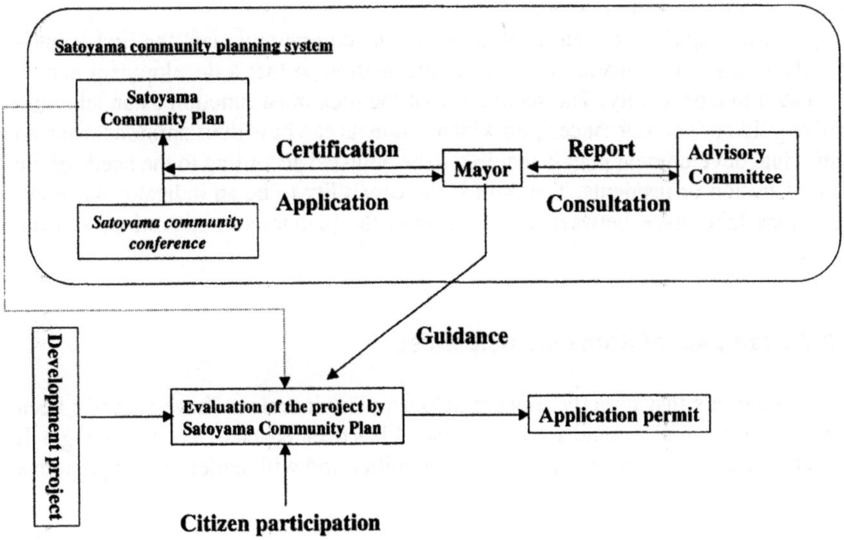

Fig. 7.6. Land use coordination system based on the Satoyama Community Plan

planning. The first is to include the whole satoyama environment in the municipalities concerned. The second is a dialogue to be used to draw up a plan for every satoyama environment, because natural and social conditions vary among municipalities. It is likely that a framework structure could be built on the basis of the first dialogue, with a second dialogue being incorporated thereafter. A model that presupposes only the second dialogue can be seen in Figure 7.6. The following discussion presu 3 pposes that establishment of a satoyama community is regulated by municipal bylaws.

The central tenet of the establishment of a satoyama community is that residents will live there in the end. A satoyama community conference is established through the participation of a certain percentage of residents (e.g., 80% of landowners; the treatment of non-local landowners will become a problem). The conference will draw up a satoyama community plan. Satoyama environments can be divided, for example, into woodland areas, satoyama landscape areas, agricultural reserve areas, and communal residence areas. The satoyama community plan should suit the comprehensive land-use direction and land-use standards and should be detailed enough to be used as a municipal plan. A municipal head will finally and formally approve a plan after submission has been made to a deliberative council. When enforcing certain restrictions in an area, one of the methods is to apply zoning as an administrative initiative; the other is to wait for suggestions from residents before applying zoning restrictions. Although the former method is not ruled out, the latter dialogue is preferable to enable the establishment of a satoyama community to take place with citizen participation.

As for the legal effectiveness of the satoyama community plan, the first point is that there has to be capacity to provide information so that a development can be evaluated independently. The second is that the plan must function as an administrative guide or as a reference upon which residents can base their opinions when an individual development plan is required to be adjusted according to the needs of the administration or residents. The third is the capability to be an indicator when negotiations take place between the developer, the planner/administration, and the residents.

7.2.7 Examples of Some Municipalities

The regulations that directly concern satoyama environments have recently been enacted and the policy trend seen in Chapter 1 has been enacted. The following will introduce the regulations of several municipalities and will explore the legal points and subjects.

7.2.7.1 The Satoyama Landscape Conservation Regulation of Kochi City

In Kochi City, the loss of familiar and valuable nature by the expansion of land for housing outside the urbanization promotion area has caused a great deal of concern. Because of this, and in order to advance community based planning that is in harmony with nature and well balanced with leisure amenities, Kochi City enacted the Satoyama Landscape Conservation Regulations on 1 April, 2000, by recognizing the importance of satoyama landscapes as a resource. This legislation imposes regulations by the following:

1. Securing a disaster-prevention role for the area and guaranteeing the recovery of a city's living environment.
2. Taking steps toward the conservation and recovery of a natural environment that aims at natural diversity, such as conservation of biological species and conservation of natural cycles.
3. Encouraging the conservation and recovery of a natural environment with citizen participation by regarding satoyama landscapes as places of learning and the passing down of local culture and history through the generations.

In the Kochi municipal regulations, the area for conservation has been designated as a satoyama landscape conservation area on the administration's initiative. The regulations provide that the administration must be notified when certain predesignated types of development occur in the satoyama landscape conservation area according to the standard procedure defined by the regulations. In a conservation area, a landowner and a mayor can conclude the satoyama landscape conservation agreement. The city concludes with a lease contract and installs *citizen's satoyama* in the conservation area to open it to development from the citizens.

7.2.7.2 Satoyama Community Ordinance

Sasayama City in Hyogo Prefecture enacted the Sayama City Community Planning Ordinance in 1999. Under these regulations, a local association for satoyama community planning is formed according to civic initiatives and a satoyama community plan is drawn up with the consent of a two-thirds majority of land-owners. The plan functions as a guidance standard on development taking place under other regulations in the Sasayama City Community Planning Ordinance when the plan has been authorized by the mayor.

7.2.7.3 Designation of a Cohabitation Area in Kobe City

In Kobe City, although the Ordinance for More Green Areas and More Access by Citizens was enacted in 1991, this regulation was applied only to open spaces in the Urbanization Control Area. Therefore, the city felt the need to take care of land use in villages and satoyama landscapes, which are not accounted for in the regulation. The Cohabitation Area legislation was then enacted in 1996.

Essentially, this regulation is intended for the Urbanization Control Area. The cohabitation area inside this is divided into an environmental conservation zone, an agricultural conservation zone, a communal residential zone, and a specific use zone (Figure 7.7). Each zone is designated according to a variety of functions, such

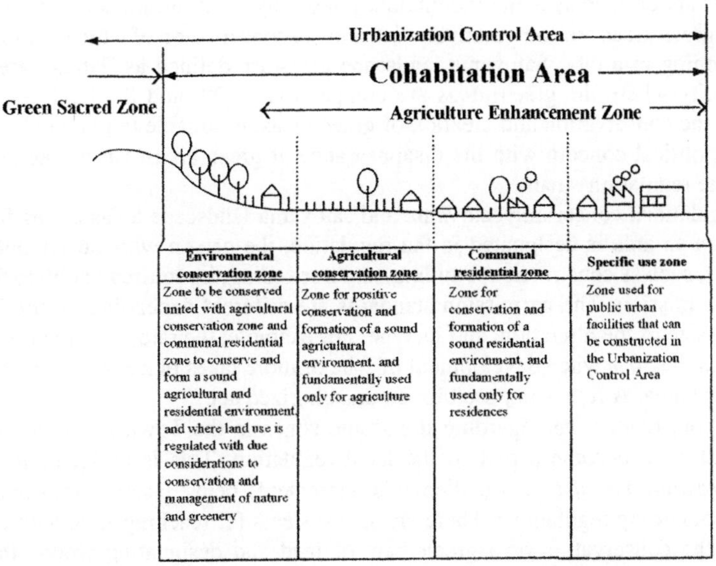

Fig. 7.7. Conceptual representation of the cohabitation zone in Kobe City

as the following:

1. The supply of fresh and safe agricultural products.
2. A comfortable living environment.
3. The conservation of a rich natural environment.
4. Provision of a place for relaxation and peacefulness.
5. Provision of a broad educational environment.
6. The passing down of traditional local culture from one generation to the next.

The mayor is notified when certain predesignated developments in the above-mentioned zones take place.

The satoyama community conference system also contains provisions in order to draw up independent rules in an area. The conference concludes the satoyama community establishment plan with the consent of a three-quarters majority of residents and receives the mayor's authorization. If the plan is authorized, the administration will support development that conforms to the regulations.

7.2.7.4 Sapporo City: Conservation and Creation of Green Areas

The Sapporo City Ordinance for Green Areas, which was enacted in 2001, provides classifications for mountainous areas, satoyama landscape areas, satoyama environment areas, residential city areas, and business city areas, and regulates for specific kinds of development in each area. According to these regulations, satoyama areas are defined as being mountainous and hilly areas around a city area in which the green areas are comparatively rich and the conservation of which is suited to the planning in a city. Satoyama landscape areas are defined as flatland areas around a city where the green areas are comparatively rich and the land use of which, by the conservation and creation of green areas, is suitable to planning in a city. The political concern with the disappearance of green areas can be seen directly in the regulation's name.

The regulations concerning satoyama and satoyama landscape areas are as follows. When an area is designated in the regulation, the person who carries out a specific act, such as construction, building, and tree felling, is required to obtain the mayor's permission. The permission standards are different depending on the location, the size of the alteration, and its type. For example, the rate of a green tract of land must be 30% when development of 5 ha or more takes place in a satoyama environment area. A fine is imposed for an unauthorized act.

In addition, regulations regarding the scenic zones in the Town Planning and Zoning Act have become a part of the local regulation. This is an example of so-called *regulations of trust* (gentlemen's agreement) being inserted into independently drawn-up regulations. There are also systems for reaching an agreement regarding the conservation of a green tract of land and designating woods that should be conserved.

7.2.7.5 Institutionalization and Utilization of a Regulation

One of the problems of such regulations is the measures that are taken when a company or a person does not comply with the standards laid out in the regulations. The surest way of satisfying standards is to prevent the company or person from obtaining permission when a proposed development does not conform with the regulations. The Sapporo City Ordinance has specified such a provision. However, such a strict measure is not generally institutionalized. Fundamentally, what the administration is doing is providing administrative guidance. Because, in the end, making a decision to follow the guidance is optional, a fine cannot be imposed. Regarding this point, for example, the name of a perpetrator can be released in some cases when the guidance is ignored, such as in the Kochi City Ordinance. To do this, indicators and standards that are used as the basis of guidance must have sufficient legitimacy.

In making the rules in an area, there is a choice whether a top–down or bottom–up style is taken. Traditionally, measures using the former were common. However, recently, as seen in the Sasayama City Ordinance and the Kobe City Ordinance, the latter measures tend to be regarded as useful. In order to utilize the versatile functions of satoyama environments, the contents of a plan need to be drawn up with active participation of the people concerned, and the bottom-up style is effective for this. Moreover, citizen participation is indispensable when making local rules that concern citizens. Unless people in an area are actively involved, institutional support for realization of a plan cannot be obtained. This system can be seen to be one that respects the autonomy of residents during a period of decentralization.

As seen previously in the Town Planning and Zoning Act revised in 2000, some guidelines for the use of municipalities were prepared. Moreover, the authority of community based planning was awarded to these municipalities so they could institutionalize the small-scale scenic zone rights for local nature protection, although the designation of the zone is limited in a city planning area. The National Diet has also claimed that satoyama landscapes may become the basis of enabling small-scale scenic zones, such as that of:

'... conserving scenic zones as much as possible even if they are only very small, such as those in satoyama landscapes or a grove of a village shrine. Municipalities which have a thorough knowledge of the actual condition of an area will now be able to decide the contents of regulations in small-scale scenic zones... ' (page 29 of the minutes of the 147th National Diet Lower House Construction Committee No. 10, 19 April, 2000).

The work of conserving satoyama environments by successfully combining the independent system of a municipality and a legal system is a major challenge in the environmental administration of a future municipality. However, caution should be exercised in this situation: certain areas may be well regulated, others, where the system does not apply, may be poorly regulated. Municipalities should discuss the implications of the strictness of regulations applied to each area.

7.2.8 The Future of Satoyama Landscape Conservation Laws

A law is the result of trying to balance benefits to society in a political way. In relation to the right of land use, the more significant the benefits that conservation is considered to bring, the more strict the land-use regulation that will be imposed. Although there might be a prohibition of certain kinds of development and a requirement of compliance to a certain standard, as well as stipulating that a procedure is carried out, in all cases these represent some kind of restriction on the rights of a landowner.

In order to enact such regulations lawfully, the effects on satoyama environments must be understood in a social context, so that even if an individual stakeholder appeals for conservation, the legislature will not automatically react to this. Moreover, even if the necessity of conservation is generally understood, in order to fully restrain a landowner's right, enough persuasive data on the importance of the social functions of satoyama environments must be presented and shared. Without this, the regulations will not be considered fair and, ultimately, legitimate. In general, the more necessary a conservation is thought to be from a social point of view, the more unavoidable the disadvantages caused by a restriction seem. In this sense, the social mission of natural scientific research is very great, indeed.

7.3 National Land Planning of Satoyama Landscapes

K. Takeuchi

7.3.1 Why does Satoyama Nature Attract Attention?

Concern about satoyama landscapes is increasing in national land planning. Satoyama environments, once regarded as unnecessary space after people began to depend on fossil fuels and chemical fertilizers, and considered suitable for large-scale development, have come to be reevaluated as an indispensable space in maintaining the affluence of current lifestyles. The step-by-step loss of familiar nature and people's nostalgic appreciation of the landscape amid an overwhelming change of values in society has come to be regarded with a great deal of concern. The nature of satoyama environments, which has remained despite encroachment from surrounding development, has recently enjoyed a renaissance, like the runner in a race who has to catch up after a handicap of one lap. With this, citizens' desire for nature is now coming to be realized and satisfied.

The charm of satoyama landscapes repeats itself in the abundance and beauty of nature that is managed by humans in an appropriate way. One way to fully enjoy satoyama nature is not by seeing it from a distance, but by enjoying it as a hands-on experience. Bequeathing satoyama landscapes to the next generation, something that was once assumed would happen forever, leads to a culture of handing down Japan's natural features to future generations. For that purpose, it is not enough that satoyama landscapes are protected in one specific area. The development of a movement for protecting and raising satoyama landscapes throughout Japan has been called for. In addition, this should be followed by a consideration of the concrete institutional and financial supports for overcoming various obstacles that make the maintenance of satoyama landscapes difficult.

Another reason why satoyama landscapes attracts attention is that they are being designated as *common*, a concept that presents an alternative to private and public, and will become important in the 21st century, the century of the environment. Satoyama landscapes have a history of being shared as common land. The resources there were managed by sharing locally. That people overuse such commons and spoil resources as a result has become known as the *tragedy of the commons* (Hardin 1968). By considering how the satoyama landscapes can be maintained as a resource to be shared locally will probably contribute to raising the consciousness toward common environmental resources in present-day society.

7.3.2 Protecting the Biodiversity of Satoyama Landscapes

Conserving satoyama landscapes means contributing to solving the problems of the global environment. In the United Nations' Convention on Biological Diversity (UNEP/CBD), every country in the world is requested to conserve its own biodiversity. Japan drew up the National Biodiversity Strategy in 1995 according to this convention (Ministers' Meeting on Global Environmental Conservation 1995). This strategy outlines long-term goals, guaranteeing a range of wildlife habitats close to human settlements, up until the middle of the 21st century. Summarizing the strategy, it calls for the conservation and continuous utilization of various animals, plants, and ecosystems locally, and also for the conservation and use of every biogeographical classification in Japan. In so doing, the strategy aims to conserve the various linkages between organisms and the proper management of large-scale protection areas by organizing area-wide networks. Although the full-scale measures for attaining such targets are still under examination, some measures related to satoyama environments have been undertaken, such as the preparation of a manual for promoting participation and cooperation of citizens for the conservation and rehabilitation of satoyama environments living together with endangered species, and special nature-oriented planning for the conservation and networking of nature.

At the 5th Conference of the Parties Convention on Biological Diversity, the principle of the ecosystem approach that should serve as the norm was adopted in order to attain three targets in the convention: the conservation of biodiversity, sustainable use, and fair, impartial distribution of the profits from the use of hereditary resources (Table 7.3). This principle is very meaningful in pointing out the importance of understanding ecosystems in terms of economics when managing the ecosystem. It also assigns priorities to the conservation of the function and structure of ecosystems. On the other hand, the former biodiversity conservation policy emphasized the management of protected areas and the protection of rare species. Moreover, it can also be noted that the necessity of flexible adaptive management is

Table 7.3 The 12 principles of the ecosystem approach (UNEP/CBD 2000)

Principle 1	The objectives of the management of land, water, and living resources are a matter of societal choices
Principle 2	Management should be decentralized to the lowest appropriate level
Principle 3	Ecosystem managers should consider the effects (actual or potential) of their activities on adjacent and other ecosystems
Principle 4	Recognizing potential gains from management, there is usually a need to understand and manage the ecosystem in an economic context. Any such ecosystem-management program should: (a) Reduce those market distortions that adversely affect biological diversity (b) Align incentives to promote biodiversity conservation and sustainable use (c) Internalize costs and benefits in the given ecosystem to the extent feasible
Principle 5	Conservation of ecosystem structure and functioning, in order to maintain ecosystem services, should be a priority target of the ecosystem approach
Principle 6	The ecosystem must be managed within the limits of its functioning
Principle 7	The ecosystem approach should be undertaken at the appropriate spatial and temporal scales
Principle 8	Recognizing the varying temporal scales and lag-effects that characterize ecosystem processes, objectives for ecosystem management should be set for the long term
Principle 9	Management must recognize that change is inevitable
Principle 10	The ecosystem approach should seek the appropriate balance between, and integration of, conservation and use of biological diversity
Principle 11	The ecosystem approach should consider all forms of relevant information, including scientific, indigenous, and local knowledge, innovations, and practices
Principle 12	The ecosystem approach should involve all relevant sectors of society and scientific disciplines

emphasized by the way in which it foresees how natural changes cannot be avoided (UNEP/CBD 2000)

By reconsidering Japan's natural conservation measures against such principles, it is clear that the existing conservation measures for areas like satoyama landscapes, where secondary nature prevails, were weak. According to the report describing the way of conserving biodiversity upon the drawing up of the new Basic Environment Plan (Ministry of the Environment 2000), it was said that the arguments did not agree when discussing in what condition Japan's secondary nature should be maintained. Based on this, it was supposed that a prioritization of measures would be needed for every natural entity to be considered as a cluster and every process that could be considered to be a sequence, instead of trying to accord the aim of the conservation with a target related to the quantitative and qualitative decrease of biodiversity in secondary nature. Furthermore, it was thought that a selection of the management type would be needed by evaluating various functions and values of secondary nature, apart from those of biodiversity. As well as this, in future, the labor required would need to be evaluated in order that the functions and values for every area are best shown.

In terms of this, supposing that national-scale satoyama landscape management encounters labor or financial difficulties, decision making is left to each area so that satoyama environments can be maintained at a level for which local residents are willing to pay. However, there may also be difficulties in understanding that the management of satoyama landscapes is public work, because these management practices are often concentrated on artificial construction, and, moreover, that such actions will weaken the financial base of the country or local municipalities. Furthermore, it is a problem that neither the country nor a local municipality cannot take a wider view of the future of satoyama environments. It is thought that organizing plans about future satoyama conservation at a national level, by recognizing the difficulty of finding a solution to the various problems, will need to occur when considering that the conservation of biodiversity is an important policy subject in Japan.

7.3.3 Satoyama Conservation in National Land Planning

The emergence of a positive attitude toward the conservation of the nature of satoyama landscapes in terms of national land planning can be noted. In the Grand Design of National Lands in the 21st Century, which was the new Comprehensive National Land Development Plan drawn up in 1998, it is stated that it is necessary to maintain secondary nature through agriculture, woodlandry, and fishery, to maintain satoyama landscapes through citizens' cooperation, and, moreover, to create woods and waterside grounds actively as part of public works projects, as well as conserving the outstanding natural environment, such as national parks. This would be in order to respond to the needs of nature-oriented people and the voice of people that call for the conservation of biodiversity (National Land Agency 1998). One of the main points here is that conservation of satoyama environments is indivisible from agriculture and forestry at the national level. In short, how to connect agriculture and forestry to local development is the focal point.

Considering this point, the new Comprehensive National Land Development Plan offers one proposal. That is, to place a local city and its surrounding rural area as the country's *new frontier* in the 21st century and to heighten the charm of residence there. This strategy is called *creation of residential area full of natural environment* and demonstrates the policy for maintaining minor cities and their surrounding rural areas as residential areas, rich in nature. This strategy shows the importance of forming plans that unite a minor city and its surrounding rural area. This has a relation with the philosophy behind Howard's garden city theory (Howard 1902), which considers that an ideal garden city is made up by the marriage of city and a farming village. It can be said here that agriculture and forestry are not only primary industries, but that they should be reconsidered to be an environment that has versatile functions and values. Therefore, that the functions and values of satoyama nature should be properly evaluated.

It is clear that satoyama environments are attractive for city residents who see nature on the outskirts of their city. However, similar qualities can also be found in rural areas that are more remotely situated, such as mountainous areas. Leisure activity in rural areas, known as *green tourism*, has grown recently. A farm village vacation is already a common activity and farmhouses that provide beds and meals also prosper in European countries, such as Britain, Germany, and France. Similarly, in Japan, the number of people spending their holidays in rural areas is increasing steadily. Unfamiliar experiences are respected more than extraordinary experiences. People who get tired of working in an artificial environment enjoy another kind of everyday life.

Of course, financial profit alone from these leisure activities cannot maintain residential areas full of natural environments. It is necessary to establish various businesses that enable people to reside in the area. However, this must be done without spoiling the charm of satoyama nature, such as welcoming old forms of polluting industry. Instead, a business in which advanced knowledge is concentrated or a welfare business corresponding to ageing in the area might be suitable. Although agriculture and forestry are unstable alone, they could be a potential success if united with processing, distribution, and sales, and are further connected to green tourism.

7.3.4 The Restoration of Lost Nature

Big cities are entering a time in which disorderly expansion is controlled and ways of improving the current city environment are being considered. The shift from an urbanizing society to a urbanized society is at last coming to Japan. The problems of disorderly urbanization must be solved. Lost nature must be regenerated and the city must be restored to the condition of a healthy ecosystem with a link to the preservation of satoyama landscapes. This has already been demonstrated in various river restoration projects. Satoyama landscapes can also be restored through tree planting in vacant lots left by city redevelopment, emphasizing the maintenance of city parks as places of secondary nature.

An ecological network philosophy is necessary. In satoyama natural areas on the outskirts of a big city, fragmentation of the ecosystem is occurring, along with a quantitative reduction in nature, as described in Chapter 2. Such fragmentation represents a major threat to biodiversity by obstructing the movement of organisms. In The Netherlands, where a high degree of artificial landscape exists, the national environmental strategy aims to solve such fragmentation. Although Japan has not come to develop environmental policy that solves fragmentation as of yet, programs that restore nature through public works are becoming increasingly common.

Mitigation in the United States has attracted attention as a technique that stops the quantitative reduction of nature and the fragmentation that accompanies development. Mitigation means environmental relief, and mandates the avoidance and mitigation of adverse influences on nature and the restoration of nature when development is unavoidable. Mitigation has been applied, for example, in the conservation of wetlands. It lays down the requirement that the restored nature is more than the lost nature based on the principle of no net loss. The balance of lost nature and nature to be restored is objectively and quantitatively indicated by the Habitat Evaluation Procedure (HEP), which decides on whether the compensatory measure is appropriate (Figure 7.8; Tanaka 1999).

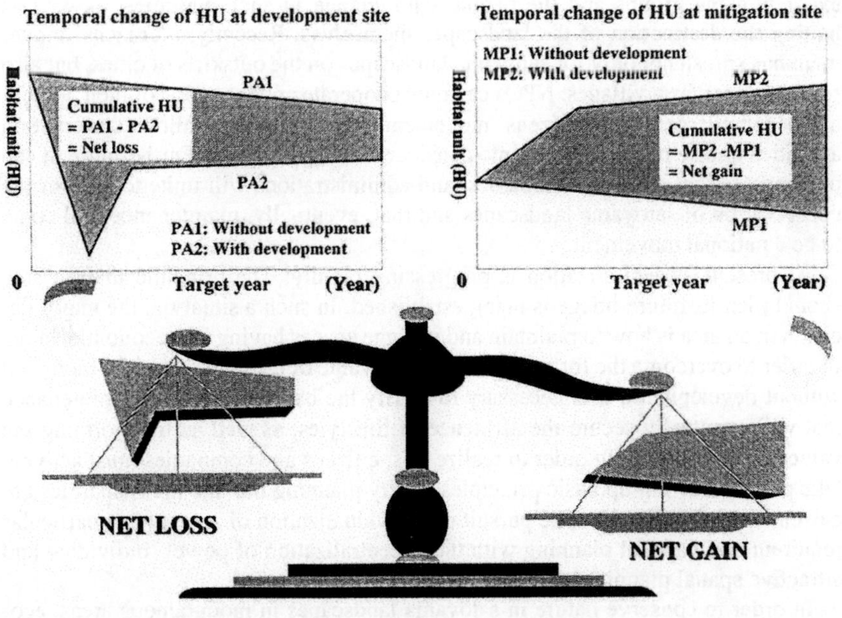

Fig. 7.8. Conceptual representation of the Habitat Evaluation Procedure (HEP). (After Tanaka 1999)

In the United States, a method that uses a mitigation bank can be utilized, whereby the restored nature can also be effectively used in forming an ecosystem network. In this case, nature is restored as a cluster and guidance in taking alternative measures in the area is also considered (Brown and Lant 1999). This process aims to effectively save the natural place rather than restoring it afterwards. Moreover, in Germany, the Federal Construction Law was recently reformed and a more advanced method is used, in which places where nature that is to be restored to account for destruction caused by other developments is specified beforehand in a planning map as ecological storage. This land-use planning has authority at the regional level of spatial planning (Wiesner 1998).

In Japan, examination of the environmental conservation measures accompanied by development can also be seen in the Basic Matters Related to Environmental Impact Assessment Law (Environment Agency of Japan 1997). Priority can be given to development that avoids adverse influences on the environment or that includes adequate, mitigating environmental conservation measures; the examination of which compensatory measure to include will be made on a case-by-case basis. Such ways of thinking are necessary not only for environmental assessment when development occurs, but also at various planning decision stages.

7.3.5 Satoyama Landscape Management Through Participation

The greatest effect in the conservation of satoyama landscapes can be obtained by expanding the groups and the people who engage in such activities, as well as halting the destruction of the landscapes themselves. Recently, a citizens' movement has spread not only for satoyama landscapes on the outskirts of cities, but also for those near farm villages. NPOs can now cooperate and the scope of their activity has become larger. Such citizens' movements, by cooperating, allow conservation activities to take place that take into consideration a wider area. Furthermore, it can be expected that citizens, companies, and administrations will unite to aim for the conservation of satoyama landscapes and that, eventually, maintenance will come to be a national movement.

At present, decentralization is progressing rapidly. The principle that an area should plan its future image is being established. In such a situation, the major decision in an area is how to maintain and manage spaces having low economic value. In order to overcome the former belief that the value of the land cannot be increased without development, it is necessary to clarify the basic principle of maintenance that will eventually secure the affluence of lifestyles, as well as transforming the values of the affluent. In order to realize this, citizens and companies must actively take part in drawing up basic principles of city planning that the municipalities are currently responsible for. The pursuit of individualization of an area is a particular requirement in spatial planning with the decentralization of power. Individual and attractive spatial planning also leads to pride in an area.

In order to conserve nature in satoyama landscapes in mountainous areas, economic difficulties will be encountered. In such areas, there is a system of direct income payment, in which compensation is paid to a certain extent to farmers that maintain various environmental functions rather than concentrating on agricultural production. Although it is necessary to discern future transitions in order to establish how much of an effect such direct income payments will have, security alone is not a sufficient condition for protecting satoyama landscapes. What is needed is a system that enables people to live and conserve the reality rather than giving the illusion of security.

The movement toward farm systematization, including the establishment of an incorporated company, has also attracted attention. The number of cases in which young laborers have been successfully attracted to employment is already increasing in forestry through marked improvements in labor conditions following the establishment of a company organization. Moreover, there are also cases in which profits are significantly increased by combining forestry with other associated businesses. It is to be expected, therefore, that the present serious situation can also be eased by such systematization in agriculture. It is to be expected that this will lead to a more positive agricultural activation policy by allowing a mixture of subsidies and direct income payments. Simultaneously, unification of forestry and agriculture can also efficiently manage satoyama landscapes in the face of a complicated natural condition.

7.3.6 Long-term Strategy for Conservation

Conservation of satoyama landscapes must not only be a short-term strategy, but also one that takes a long-term view. The greatest difference between these two strategies is probably the change in the socioeconomic environment. If the growth of a big city slows, qualitative reduction, such as abandonment of management activities, will become a bigger problem than a quantitative reduction in the area of satoyama landscapes on the outskirts of a big city. In such an area, it is considered that a natural conservation policy that is similar to that in mountainous natural regions would be needed. In this case, the designation of a Nature Park will be necessary, as well as expecting common use and management by citizens, especially in the conservation of areas where many rare species, the conservation of which has become necessary, are distributed.

There are two possible futures for satoyama landscapes in rural areas. One is to discover new benefits as a repository of biomass resource through rapid progress in developing alternative energy sources that replace fossil fuels. As described in Chapter 5, the technical development in energy use of wood resources has seen some progress. Japan is behind in the development of such uses, despite being blessed with one of the world's leading wood resources. If production costs can be reduced drastically by changing social awareness and progress can be made in the technical development of the use of Japanese wood resources for energy, satoyama landscapes could be reappraised as the treasury of biomass. Assuming that biomass use increases, the regulation of such a function and other functions including biodiversity will become an important subject of study.

The other possibility is that the world's wood and food supply will become ever tighter, and the import of wood and agricultural products will slow this progress by aggravating global environmental issues and draining biomass resources in developing countries. In such a case, the national wood and food supply base should be strengthened. In addition, policies that improve productivity without spoiling the richness of the nature in satoyama landscapes, once it has been widely recognized among the general public today, will be needed. In any event, the present condition of continuing dependence on a variety of overseas biomass resources in spite of a very high national agriculture and forestry production potential should be called into question.

References

(J-E) in Japanese with an English abstract
(J-G) in Japanese with a German abstract
(J) in Japanese

Abbot PG, Lowore JD (1999) : Characteristics and management potential of some indigenous firewood species in Malawi. Forest Ecology and Management 119(1-3): 111-121

Abe R (1999) : *Kyoto-shi shizen fukei jorei ni tsuite* (City ordinance governing the conservation of natural scenery). Journal of the Japanese Institute of Landscape Architecture 63: 160-161 (J)

Amorini E, Manetti MC, Turchetti T, Sansotta A, Villani F (2001): Impact of silviculture system on *Cryphonectria parasitica* incidence and on genetic variability in a chestnut coppice in central Italy. Forest Ecology and Management 142(1-3): 19-31

Anderson MK (1999): The fire, pruning, and coppice management of temperate ecosystems for basketry material by California Indian tribes. Human Ecology 27(1): 79-113

Azuma A, Takeuchi K (1999): Relationships between population density of frogs and environmental conditions in *yatsu*-habitat. Journal of the Japanese Institute of Landscape Architecture 63: 573-576 (J-E)

Azuma A, Takeuchi K, Tsunekawa A (1998): Behavior of gray-faced buzzard eagles and their habitat use in *yatsu*-environment. Papers on Environmental Information Science 12: 239-244 (J-E)

Azuma A, Tokita K, Takeuchi K, Tsunekawa A (1999): Land condition of gray-faced buzzard, *Butastur indicus*, habitats in watershed of Tega marsh, Chiba Prefecture. Journal of Rural Planning Association 1: 253-258 (J-E)

Barkham JP (1992): The effect of coppicing and neglect on the performance of the perennial ground flora. In: Buckley GP (ed) Ecology and management of coppice woodlands. Chapman & Hall, London, pp. 115-146.

Bennett G (ed) (1991): Towards a European ecological network. Institute for European Environmental Policy, The Netherlands

Booker J, Tittensor R (1992): Coppicing for nature conservation: - the practical reality. In: Buckley GP (ed) Ecology and management of coppice woodlands. Chapman & Hall, London, pp. 299-305

Brookfield H, Byron Y (eds) (1993): South-East Asia's environmental future: The search for sustainability. United Nations University Press, Tokyo

Brown PH, Lant CL (1999): The effect of wetland mitigation banking on the achievement of no-net-loss. Environmental Management 23: 333-345

Buckley GP (ed) (1992): Ecology and management of coppice woodlands. Chapman & Hall, London

Bureau of Environmental Protection, Tokyo Metropolis (1998): *Toukyou-to no hogo-jou jyuyou-na yasei-seibutsu-shu* (Important wildlife species for the protection in Tokyo Metropolis). Tokyo Metropolis, Tokyo (J)

Cutini A (2001): New management options in chestnut coppices: an evaluation on ecological basis. Forest Ecology and Management 141(3): 165-174

Dubuission X, Sintzoff I (1998): Energy and CO_2 balances in different power generation routes using wood fuel from short rotation coppice. Biomass and Bioenergy 15(4-5): 379-390

Enari T (2000): Vision of maintaining satoyama by urban citizens. Journal of the Japanese Institute of Landscape Architecture 63: 186-189 (J-E)

Enomoto T (1999): *Satoyama no atarashii kachikan wo kangaeru* (Considering new values of satoyama). Gotomaki-dayori, 115: p 1 (J)

Environment Agency of Japan (1994): *Shin kankyo kihon keikaku* (The New Basic Environment Plan). Environment Agency, Tokyo (J)

Environment Agency of Japan (1997) : *Kankyo-eikyo-hyouka-hou no kitei ni motozuku kihon teki-jikou* (Basic matters related to the Environmental Impact Assessment Law). Environment Agency, Tokyo (J)

Environment Agency of Japan (2000): Threatened Wildlife of Japan: –Red Data Book 2nd ed, Vol. 8, Vascular Plants 2000. Environment Agency of Japan (J)

Environment Agency of Japan, 'Satochi' Research Group (1996): *'Satochi' kara no henkaku* (Reform from 'satochi'). Jiji-tsushin-sha, Tokyo (J)

Environmental Laboratory, US Army Corps of Engineers (1987): Corps of engineers wetlands delineation manual, (http://www.saj.usace.army.mil/permit/document/87manual.pdf)

Evans MN, Barkham JP (1992): Coppicing and natural disturbance in temperate woodlands: - a review. In: Buckley GP (ed) Ecology and management of coppice woodlands. Chapman & Hall, London, pp. 79-98

Fleming TH (1991): Fruiting plant-frugivore mutualism: The evolutionary theater and ecological play. In: Price PW, Lewinshoh TM (eds) Plant-animal interactions: Evolutionary ecology in tropical and temperate regions. John Wiley & Sons, New York, pp. 119-144

Forest Agency of Japan (1978): *Satoyama chiiki kaihatsu hozen keikaku houkokusho* (Report for development and conservation plannings of woodland in satoyama landscape). Forest Agency of Japan, Tokyo (J)

Forman RTT (1995): Land mosaics –The ecology of landscape and regions. Cambridge University Press, Cambridge, UK

Forman RTT, Godron M (1986): Landscape ecology. John Wiley & Sons, New York

Fortune R. (1863): Yedo and Peking –A narrative of a journey to the capitals of Japan and China. John Murray, London

Fujioka M, Lane S J (1997): The impact of changes in irrigation practices in rice fields on frog populations of the Kanto Plain, central Japan. Ecological Research 12: 101-108

Fujiwara H (1998): *Inasaku no kigen wo saguru* (Inquiring into the origin of rice cultivation). Iwanami-shoten, Tokyo (J)

Fukamachi K, Oku H, Nakashizuka T (2001): The change of a satoyama landscape and its causality in Kamiseya, Kyoto Prefecture, Japan. Landscape Ecology 16: 703-717

Fukui A (1993): *Hishoku-shushi-sampu ni okeru doshokubutsu no sougo-kankei* (Plant-animal interactions in frugivorous seed dispersal). In: Washitani I, Ogushi T (eds) Interrelationships between plants and animals. Heibon-sha, Tokyo, pp. 222-235 (J)

Fukuyama K, Sato D (1998): *Zushi-Onoji-chiku no ryousei-rui* (The amphibia in Zushi-Onoji area). Rekikan-tushin 1: p 6 (J)

Fuller R J (1992): Effects of coppice management on woodland breeding birds. In: Buckley GP (ed) Ecology and management of coppice woodlands. Chapman & Hall, London, pp. 169-192

Goldsmith FB (1992): Coppicing: - a conservation panacea? In: Buckley GP (ed) Ecology and management of coppice woodlands. Chapman & Hall, London, pp. 306-312

Gurnell J, Hicks M, Whitbread S (1992): The effects of coppice management on small mammal populations. In: Buckley GP (ed) Ecology and management of coppice woodlands. Chapman & Hall, London, pp. 213-232

Hardin G (1968): The tragedy of the commons. Science 162: 1243-1248

Hattori T, Akamatu H, Takeda Y, Kodate S, Kamihogi A, Yamazaki H (1995): *Satoyama no genjyo to satoyama kanri.* (The present status of satoyama and its management). Man and Nature 6: 1-32 (J)

Hattori T, Kamihogi A, Kodama S, Kumadaki E, Fujii T, Takeda Y (1994): A Study on the Actual Conditions of the Fragmented Forests in Flower Town and their Conservation. Journal of the Japanese Institute of Landscape Architects 57(5): 217-222 (J-E)

Hayama Y (1994): Fundamental studies for the relationship between vegetation and avifauna in the urban forest at breeding season. Journal of the Japanese Institute of Landscape Architecture 57(5): 229-234 (J-E)

Hayama Y, Takahashi R, Katsuno T (1996): The relationship between wood structure and avifauna of Higashiyamato park, Tokyo. Journal of the Japanese Institute of Landscape Architecture 59(5): 89-92 (J-E)

Higuchi H, Tsukamoto Y, Hanawa S, Takeda S. (1982): Relationship between forest areas and the number of bird species. Strix 1: 70-78 (J-E)

Himiyama Y (1995): *Kokudo riyou henka no gaiyou* (Outlines of national landuse changes). In: Nishikawa O (ed) Atlas — environmental changes in modern Japan. Asakura shoten, Tokyo, 1-16 (J)

Hong SK, Nakagoshi N, Nehira K (1993): Trends of *Pinus densiflora* populations under the traditional regimes of forest management in the rural landscapes of Korea and Japan. Annali di Botanica LI: 6-20

Howard E (1902): Garden cities of to-morrow. Swan Sonnenschein, London

Huston MA (1994): Biological diversity: the coexistence of species on changing landscapes. Cambridge University Press, New York

Ichinose T, Katoh K (1994): The factors that influenced bird communities on the fragmented woodlots in the Tokorozawa City, Saitama Prefecture. Journal of the Japanese Institute of Landscape Architecture 57(5): 235-240 (J-E)

Ichinose T, Katoh K (1996): A relationship between bird communities and vegetation structure on isolated woodlots in wintering season in Tokorozawa City,

Saitama Pref., Japan. Journal of the Japanese Institute of Landscape Architecture 59(5): 73-76 (J-E)

Ichinose T, Katoh K (1998): Factors influencing bird distribution among isolated woodlots on a heterogeneous landscape in Saitama Pref., Japan. Oekologia (Bratislava) 17: 298-310

Ichinose T, Katoh K (1999): Relationship in species composition between vegetation and bird communities in remnant woodlots on Musashino Uplands and Sayama Hills. Journal of the Japanese Institute of Landscape Architecture 62: 577-580 (J-E)

Ide M (1992): Ecological studies on the rural landscape patterns for the conservation of biotopes. Bulletin of the Laboratory of Landscape Architecture and Science, The University of Tokyo 11: pp 1-120 (J-E)

Ikeno S (1993): *Ibaraki-ken no genjo to mirai: – anazuka-oike no sashiba wo chushin toshite* (Present and future of Ibaraki Prefecture: – with special reference to Gray-faced Buzzard eagles in Anazuka-oike). Anazuka-no-sizen-to-rekisi-no-kai, Ibaraki (J)

Inui T (1982): On the Significant utilization of the flatland forests in agriculture in the western suburbs of Tokyo. Geographical Review of Japan 55: 549-565 (J-E)

Ishii M, Ueda K, Sigematu (1993) *Satoyama no shizen wo mamoru* (Preserving nature of satoyama). Tsukiji-shokan, Tokyo (J)

Kagawa T (1999): *Satoyama no hozen-kanri to amenity* (Conservation management and amenity in satoyama). In: Japanese Institute of Landscape Architecture (ed) Landscape Ecology. Gihoudou-shuppan, Tokyo, 72-83. (J)

Kameyama A (ed) (1996): *Zohki-bayashi no shokusei-kanri* (Vegetation management of coppice). Soft-science, Tokyo (J)

Kameyama A et al. (1994): *Ikimono-gijutsu toshiteno zoen* (6) (The JILA periodical meeting (spring 1993) working session 'Landscape architecture as technology for lives'). Journal of the Japanese Institute of Landscape Architecture 57: 279-286 (J)

Kamiya M (1980): *Ikigai ni tsuite* (On life worth living). Misuzu-shobo, Tokyo (J)

Kasahara Y (1951): *Suiden-zassou no chiriteki-bunpu to hassei-do* （Geographical distributions and degrees of occurrences of paddy weeds） . Nougaku-kenkyu 39(4): 143-154. (J)

Kasahara Y (1954): *Jyuurai no kenkyu no teisei-hoi to kaikon-chi no ichijiteki zassou narabini chihouteki zassou ni tsuite* (Corrections on previous reports and on the weeds temporarily occurring in reclimed sites and local weeds). Nougaku-kenkyu 42(3): 97-113. (J)

Kato M (2001): 'SATOYAMA' and biodiversity conservation SATOYAMA as important insect habitats. Global Environmental Research, 5: 135-149

Katoh K (1996): Relationship between vegetation structure and wintering avifauna in urban openspace woods. Journal of the Japanese Institute of Landscape Architecture 59(5): 77-80 (J-E)

Kennedy AD (1998): Coppicing of *Tarconanthus camphoratus* (Compositae) as a source of sustainable fuelwood production: an example from the Laikipia Plateau, Kenya. African Journal of Ecology 36(2): 148-158

Kimura M, Ito Y (1960): *Shinden sonraku* (Rural communities in newly-developed paddy field area). Bungadou, Tokyo (J)

Kitagawa Y, Yamada S (2001): *Tama-kyuryou niokeru yatoda no shokubutsu-so* (The flora of "*yatoda*", the rice paddy in small valley, on the Tama Hills). Natural History Report of Kanagawa 22: 13-18 (J)

Komazawa A, Kuramoto N (2001): *Satoyama-gata-kouen ni okeru volunteer katsudou to coordinator no shokunou ni kansuru kentou* (Examination about the function of the volunteer activity and coordinator in Satoyama park). Technical Reports of Landscape Architecture 1: 136-137

Kominami Y (1993): *Chorui no kajitsushoku to shushi-sampu* (Frugivory of birds and seed dispersal, In: Washitani I, Ogushi T (eds) Interrelationships between plants and animals. Heibon-sha, Tokyo, pp. 207-221 (J)

Konvika M, Kuras T (1999): Population structure, behaviour and selection of oviposition sites of an endangered butterfly, *Parnassius mnemosyne*, in Litovelske Pomoravi, Czech Republic. Journal of Insect Conservation 3(3): 211-223

Kuramoto N (1994): *Tama-kyuryou no toshi-kouen niokeru zouki-bayashi no shimin-sanka-gata shokusei-kanri* (Coppice management participated by citizen on municipal park in Tama Hills). Ringyo-Gijutu 629: 38-39 (J)

Kuramoto N (1996): *Zouki-bayashi no shimin-sanka-gata shokusei-kanri* (Coppice management participated by citizen). In: Kameyama A (ed) *Zouki-bayashi no shokusei-kanri* (Vegetation management of coppice). Soft-science, Tokyo, pp 223-229 (J)

Kuramoto N, Imoto I (1998): *Satoyama coordinator to shiteno zouenka no yakuwari* (Landscape architect as a coordinator in Satoyama). Journal of the Japanese Institute of Landscape Architecture 61: 322-324 (J-E)

Kuramoto N, Otsuka Y, Yagi M (2001): *Satoyama ni okeru dentouteki na kanri-shuhou no chiikisa to hyoujun shiyousho no arikata* (Regionally different and standard method of traditional management in satoyama). Technical Reports of Landscape Architecture 1: 4-5 (J)

Kurita H, Yokohari M (2001): Ecological potentials of golf courses on hills for the conservation of satoyama landscapes. Journal of the Japanese Institute of Landscape Architecture 64: 589-594 (J-E)

Lam TBT, Hirose, J, Iiyama K, Nakano J (1986): Total utilisation of lignocellulosic materials. Part 2. Preparation of metal chelating materials from carboxymathylated refiner mechanical pulp. Mokuzai Gakkaishi 32: 44-50

Lam TBT, Iiyama K, Nakano J (1985): Total utilisation of lignocellulosic materials. Part 1. Reparation and properties of liquid absorbents from carboxymethylated refiner mechanical pulp. Mokuzai Gakkaishi 31: 752-760

Landauer K, Brazil M (eds) (1990): Tropical home gardens. United Nations University Press, Tokyo

Lane JS, Fujioka M (1998): The impact of changes in irrigation practices on the distribution of foraging egrets and herons (Ardeidae) in the rice fields of central Japan. Biological Conservation 83: 221-230

Logli F, Joffe R (2001): Individual variability as related to stand structure and soil condition in a Mediterranean oak coppice. Forest Ecology and Management 142(1-3): 53-63

Makino A, Morimoto J, Shibata S, Osawa N, Nakanishi A (2002): *Toshi kinkou nijirin ni okeru shomenseki-bassai chokugo no mokuhon shokusei no tayosei no henka* (Changes in woody plant diversity soon after tree cutting in a small area of suburban secondary forest). Journal of the Japanese Society of Revegetation Technology 28(1) : (in press) (J)

Masubuti K, Kaminishi T (1997): Vegetation in the early Meiji era in Tama Hills, the South Kanto, Bulletin of the Kawasaki Municipal Science Museum for Youth 8: 1-24 (J)

Matthews RW (2001): Modelling of energy and carbon budgets of wood fuel coppice systems. Biomass and Bioenergy 21(1): 1-19

Ministers' Meeting on Global Environmental Conservation (1995): National Strategy for Biodiversity Conservation. Government of Japan, Tokyo (J)

Ministry of Agriculture, Nature Management and Fisheries of the Netherlands (1990): Nature policy plan of the Netherlands. Den Haag

Ministry of the Environment (2000): *Seibutsu-tayousei no kentou-team houkokusho* (Report of research team on biodiversity conservation). Ministry of the Environment, Government of Japan, Tokyo (J)

Ministry of the Environment (2002): *Shin seibutsu-tayousei kokka senryaku* (New national biodiversity strategy of Japan). Ministry of the Environment, Government of Japan, Tokyo., (http://www.biodic.go.jp/nbsap.html) (J-E)

Miyawaki A, Aoto K, Aoto Y, Fujiwara K, Furuya M, Hamada H, Harada H, Inoue K, Kajiyama M, Kawamura Y, Kimura I, Kobayashi Y, Matsuura S, Ohno K, Okuda S, Sasaki Y, Shinoda A, Suzuki K, Suzuki S, Takeda M, Tohma H, Yamaguchi T (1972): *Kanagawa-ken no genzon shokusei* (Realen Vegetation der Präfektur Kanagawa), the Board of Education of Kanagawa Pref., Yokohama (J-G)

Miyawaki A, Okuda S, Okutomi K, Tsuji S (1974): Map of Actual Vegetation of Tokyo Metropolis. Tokyo Metropolitan Government, Tokyo (J)

Miyawaki A, Sasaki Y, Iino K (1976): Phytosociological study on Plateau Hiki, Pref. Saitama. The Yokohama Phytosociological Society, Yokohama (J)

Morimoto J (2000): Studies on the dynamic changes, the conservation and the restoration of the native rhododendron colonies in the forest around Kyoto City. PhD thesis, Kyoto University, Kyoto, Japan, pp. 1-106 (J-E)

Morimoto J, Yoshida H (1999a): Estimation of the growing rate of the crowns and the prospective structure in *Rhododendron reticulatum.* Journal of the Japanese Institute of Landscape Architecture 62: 529-532 (J-E)

Morimoto J, Yoshida H (1999b): The mechanism for regulating the number of flowers at the shoot level and the estimation of the number of flowers using transition matrices in *Rhododendron reticulatum*. Journal of Japanese Forestry Society 81: 203-209 (J-E)

Morimoto J, Yoshida H (2000): Ecological evaluation of secondary forests by the survival of the native rhododendrons. In: Japanese National Committee for International Federation of Landscape Architects (ed) Regional landscape architecture strategy in a global century, IFLA Eastern Regional Conference 2000, Tokyo, pp. 70-76 (E)

Morimoto Y, Ueno H, Matusita Y (2001): Origin and changes of Shinodayama wetland with special references to conservation. Bulletin of the International Society of Landscape Ecology, Japanese Chapter 6: 39–42 (J)

Moriyama H (1988): *Shizen wo mamoru toha douiu kotoka* (On nature conservation). Nousangyoson–bunka-kyoukai, Tokyo (J)

Moriyama H (1997): *Mura no shizen wo ikasu* (Reviving rural nature). Iwanami-shoten, Tokyo (J)

Nature Conservation Society of Japan, World Wide Fund for Nature Japan (1989): *Wagakuni niokeru hogojou juuyouna shokubutsushu no genjou* (States of plant species important for conservation in Japan). NACS-J & WWFJ, Tokyo (J)

Nakagawa S (1996a): *Kanagawaken-nai niokeru shimin-sanka no mori-zukuri niokeru shokusei-kanri* (Vegetation management in citizen participated plantation in Kanagawa Prefecture). In: the Japanese Environmental Education Society (ed) Proceedings of the Seventh Meeting of the Japanese Environmental Education Society, the Japanese Environmental Education Society, Kobe, p. 119 (J)

Nakagawa S (1996b): *Saisei no zoukibayashi kara* (From coppice woodland revitalized). Sourinsha, Tokyo (J)

Nakashizuka T, Iida S (1996): *Zokibayashi no shu-tayousei* (Species diversity in coppice). In: Kameyama A (ed) Vegetation management in coppice. Soft-science, Tokyo, pp. 17-24 (J)

National Land Agency (1998): *21-seiki no kokudo no grand-design* (Grand design of national lands in the 21st century). National Land Agency, Government of Japan, Tokyo

National Statistic Office, Department of Energy, Republic of the Philippines (1989): Household energy consumption survey (HECS), National Statistic Office, Department of Energy, Manila

National Statistic Office, Department of Energy, Republic of the Philippines (1995): Household energy consumption survey (HECS), National Statistic Office, Department of Energy, Manila

Natsuhara Y, Imai C (1999): Prediction of species richness of breeding birds by landscape-level factors of urban woods in Osaka pref., Japan. Biodiversity and Conservation 8:239-253

Nousei Chousa Iinkai (Agricultural Policy Research Committee) (1977): *Nihon nougyou kiso toukei* (Basic statistics of Japanese agriculture). Nourin-toukei-kyoukai, Tokyo (J)

Ogino Y (1998): *Shinmei-yato de kirokusareta tonbo* (The dragonflies in Zushi-Onoji area) Rekikan-tushin 1: p 7 (J)

Ogura J (1992): *Ezu kara yomitoku hito to keikan no rekishi* (History of human and landscape decoded from pictures). Yuzankaku, Tokyo (J)

Ogura J (1994): Forests of the Kanto region in the 1880's. Journal of Japanese Institute of Landscape Architecture 57(5): 79-84 (J-E)

Ogura J (1996): *Shokusei kara yomu Nippon-jin no kurashi* (Japanese life read from vegetation). Yuzankaku, Tokyo (J)

Okazaki A (1970): *Shinrin fuuchi to recreation* (Scenic beauty and recreation of forest) Nihon-ringyou-chousakai, Tokyo (J)

Ono K, Meguro A, Iiyama K (2002): Carbon circulation and functions for environment of urban green spaces in Metro Manila. In: Ohmachi T et al. (eds) Strategies for environmental rehabilitation in metro Manila. University of the Philippines Press, Deliman, Philippines, 173-181

Osawa S, Katsuno T (1997): Analysis and evaluations of yato scenery by frogs' eyes in urban area. Papers on Environmental Information Science 11: 237-242 (J-E)

Osawa S, Katsuno T (1998): Discussion of peculiarity in yato (paddy field at bottomland with sideslope) and conservation of frogs. Journal of the Japanese Institute of Landscape Architecture 61: 529-534 (J-E)

Peterken GF (1992): Coppices in the lowland landscape. In: Buckley GP (ed) Ecology and management of coppice woodlands. Chapman & Hall, London, 4-16

Research Group for Red Data Book in Kinki Region (1995): *Kinki-chihou no hogojou juuyou-na shokubutsu -red data book Kinki* (States of species important for conservation in Kinki region). Kansai-sizenhogo-kikou, Osaka (J)

Rosenqvist H, Aronsson P, Hasselgren K, Perttu K (1997): Economics of using municipal water irrigation of willow coppice crops. Biomass and Bioenergy 12(1): 1-8

Rydberg D (2000): Initial sprouting, growth and mortality of European aspen and birch after selective coppicing in central Sweden. Forest Ecology and Management 130(1-3): 27-35

Sage RB (1998): Short rotation coppice for energy: towards ecological guidelines. Biomass and Bioenergy 15(1): 39-47

Sakuragaoka-Koen-Zokibayasi-Volunteers (Sakuragaoka Park Coppice Volunteers) (2001): *Zouki-bayashi no saisei: – volunteer jyuu-nen no ayumi* (Reclamation of coppice woodland: – ten years history of volunteers). Sakuragaoka Park Coppice Volunteers and Sakuragaoka Park Office, Tokyo (J)

Scholes H (1998): Can energy crops become a realistic CO_2 mitigation option in south west England? Biomass and Bioenergy 15(4-5): 333-344

Senelwa K, Sims REH (1999): Fuel characteristics of short rotation forest biomass. Biomass and Bioenergy 17(2): p 127

Shidei T (2000): *Satoyama no koto* (Miscellaneous notes of farm forest). Bulletin of Kansai Organization for Nature Conservation 22(1): 71-77 (J)

Shigematsu T (1985): The effect of improvement of light condition on flowering of wild rhododendron in secondary forest floor. Journal of the Japanese Institute of Landscape Architecture 48(5): 151-156 (J-E)

Shinjo T, Hoshino M (1989): Studies on the introduction and cultivation of the tropical pasture grasses in South-Western Islands, Japan (I). On the comparison of five tropical pasture grasses. Bulletin of the College of Agriculture, University of the Ryukyus 36: 131-136 (J-E)

Shishizuka Nature and History Club (1995a): *Kieru sato no shizen* (Disappearing satoyama nature). In: Shishizuka Nature and History Club (ed) Reports of the 2nd Satoyama Summit Symposium. Shishizuka Nature and History Club, Tsuchiura, pp. 206 (J)

Shishizuka Nature and History Club (1995b): *Shishizuka-ooike chiiki shizen-kankyo chosa hokokusho* (Reports of the investigation on the natural environment of Shishizuka-ooike area). Shishizuka Nature and History Club, Tsuchiura, pp.224 (J)

Shishizuka Nature and History Club (1999): *Satoyama no kurashi* (Life in the satoyama). Shishizuka Tsuchiura Municipality, Shishizuka Nature and History Club, Ibaraki (J)

Sims REH, Maiava TG, Bullock BT (2001): Short rotation coppice tree species selection for woody biomass production in New Zealand. Biomass and Bioenergy 20: 329–335

Smith RD, Ammann A, Bartoldus C, Brinson MM (1995): An approach for assessing wetland functions using hydrogeomorphic classification, reference wetlands, and functional indices. Technical Report WRP-DE-9, US Army Corps of Engineers Waterways Experiment Station, Vicksburg, MS

Sprague DS, Goto T, Moriyama H (2000): GIS analysis using the rapid survey map of traditional agricultural land use in the early Meiji era. Journal of the Japanese Institute of Landscape Architecture 63: 771-774 (J-E)

Stephens W, Tyrrel SF, Tiberghien JE (2000): Irrigating short rotation coppice with landfill leachate: constraints to productivity due to chloride. Bioresource Technology 75(3): 227-229

Suzuki S, Shintani H, Park SY, Saito K, Laemsak N, Okuma M, Iiyama K (1998): Preparation of binderless boards from steam exploded pulps of oil palm (*Elaeis guneensis* Jaxq.) fronds and structural characteristics of lignin and wall polysaccharides in steam exploded pulps to be discussed for self-bindings. Holzforschung 52: 417-426

Takahashi K, Hatano G (1990): *Shougai hattatsu no shinrigaku* (Psychology on development of life). Iwanami-shoten, Tokyo

Takeuchi K (1991): *Chiiki no seitaigaku* (Regional ecology). Asakura-shoten, Tokyo (J)

Tamagawa Village Board of Education (1995): Tamagawa village flora. Tamagawa Village, Saitama Prefecture (J)

Tamura S (1994): Vegetation of magusaba, and the secondary forests origins in magusaba vegetation. Bulletin of the Saitama Museum of Natural History 12: 73–82 (J)

Tamura T (1981): Multiscale landform classification study in the hills of Japan: Part II. Application of the multiscale landform classification system to pure geomorphological studies of the hills of Japan. The Scientific Reports of the Tohoku University, 7th Series (Geography) 31(2): 85-154

Tamura T, Yamamoto H, Yoshioka S (1983): A national summary of recent trends of large-scale landform transformation in Japan Geographical Review of Japan 56: 223-242 (J-E)

Tanaka A (1999): A compensatory mitigation case study in the United States and its prospects in Japan. Journal of the Japanese Institute of Landscape Architecture 62: 581-586 (J-E)

Tange T, Norisada M, Egami H, Niiyama K, Abd Rahman K (2000): Responses of *Shorea curtisii* and *Shorea leprosula* seedlings to short-term flooding. Journal of Tropical Forest Science. 12: 414-417

Tokoro M (1980): *Kinsei-ringyoshi no kenkyu* (Research on the modern history of forestry). Yoshikawa-kobunkan, Tokyo (J)

Tokyo Volunteer Center (1989): *Konnichiteki joukyouka deno volunteer katsudou ni kansuru kihonmondai kenkyu* (Studies on fundamental problem of volunteer activities in the present situation). Tokyo Volunteer Center, Tokyo (J)

Tokyo Volunteer Center (1996): *Volunteer-coordinator kenshuu-taikei to sono kangaekata* (Training system of volunteer coordinator). Tokyo Volunteer Center, Tokyo (J)

Troll C (1968): *Landschafts0kologie*. In: Tüxen R (ed): *Pflanzensoziologie und Landschafts0kologie*. Dr. Junk Verlag, Denn Haag, 1-21, 40-43

Tsuchiya M (1998): *Yokohama-shiki: – Shimin-sanka no mori-zukuri* (Plantation participated by citizen: -Yokohama methods). In: Kuramoto N, Uchijou M (eds) *Zoukibayashi wo tsukuru* (Making coppice woodland). Hyakusuisya, Tokyo, pp. 138-168 (J)

Tsuji S, Hoshino Y (1992): *Konara nijirin no rinshou kanri no henka ga shusosei to dojyo ni oyobosu eikyo* (Influences of changes in forest floor management on species composition and soil in a scarlet oak *Quercus serrata* secondary woodland). Japanese Journal of Ecology 42(2): 125-136 (J)

United Nations Environmental Program/Convention on Biological Diversity (2000): Ecosystem approach. UNEP/CBD/COP/5/23, Nairobi (Kenya), 103-109.

Uchinuma M (1997): *Zoukibayashi-sagyou wo yattemimashita* (Our exciting experience of coppicing). In: Kuramoto N, Uchijou M (eds) Making coppice woodland. Hyakusuisya, Tokyo, pp. 5-14 (J)

Ueno H (2001): Restoration ecological study on Shinodayama wetland. Masters Thesis of Graduate School of Agriculture and Biological Sciences, Osaka Prefecture University, Osaka (J-E)

Vandenhove H, Thiry Y, Gommers A, Goor F, Jossart JM, Holm E, Gaufert T, Roed J, Grebenkov A, Timofeyev S (2001): Short rotation coppice for revaluation of contaminated land. Journal of Environmental Radioactivity 56(1-2): 157-184

Wagakuni niokeru hogojou juuyou-na shokubutushu oyobi gunraku nikansuru kenkyuu-iinkai shu bunkakai (1989): *Wagakuni niokeru hogojou juuyouna shokubutsushu no genjou* (States of plant species important for conservation in Japan) NACS-J, WWFJ, Tokyo (J)

Warren MS, Thomas JA (1992): Butterfly response to coppicing. In: Buckley GP (ed) Ecology and management of coppice woodlands., Chapman & Hall, London, pp. 249-270

Washitani I (1994): *Zetsumetsu-kigushu no hanshoku/shushiseitai* (Reproductive/seed ecology of endangered plant species). Kagaku 64: 617-624 (J)

Washitani I (1998): *Sakurasou no me: Hozen-seitaigaku towa nanika?* (Primula eye: what is conservation biology?). Chijin-shokan, Tokyo (J)

Washitani I (1999): *Seibutsu-hozen no seitaigaku* (Ecology for biological conservation). Kyoritsu-shuppan, Tokyo (J)

Washitani I (2001a): Traditional sustainable ecosystem 'SATOYAMA' and biodiversity crisis in Japan: conservation ecological perspective. Global Environmental Research 5: 119-133

Washitani I (2001b): *Seitaikei wo yomigaeraseru* (Restoring ecosystems). NHK Publishing, Tokyo (J)

Washitani I, Iijima H (1999): *Yomigaere asaza saku mizube*: *Kasumigaura kara no chosen* (Restoring lakeshore ecosystem with Asaza flowers: challenges from Lake Kasumigaura). Bunichi-sogo-shuppan, Tokyo (J)

Washitani I, Suzuki K, Kato M, Ono M (1989): *Maruhanabachi handbook* (A field guide to bumblebees). Bunichi-sogo-shuppan, Tokyo (J)

Washitani I, Yahara T (1996): *Hozen-seitaigaku nyumon* (Introduction to conservation ecology). Bunichi-sogo-shuppan, Tokyo (J)

Whittaker RH (1960): Vegetation of the Siskiyou mountains, Oregon and California. Ecological Monographs 30: 279–338

Wiesner H (1998): Oekokonto: – Neues Schlagwort oder praktikables Instrument für eine nachhaltig Raumplanung. UVP- report 2/98, 118-121

Yabu S, Shigematsu T, Takahashi R (1985): Studies on site factors from the viewpoint of ecological management of *Erythronium japonicum* population. Journal of the Japanese Institute of Landscape Architects 48(5): 157-162 (J-E)

Yajima J (1954): *Nougyou to sonraku seikatsu* (Agriculture & livelihood in rural communities). In: Kiuchi S (ed) *Jinko shuraku chiri* (Population and settlement geography). Arasuka shoten, Tokyo, 30-84 (J)

Yamada S, Takeuchi K, Kitagawa Y (2000): Influence on vegetation due to mowing, plowing and surface soil paddling in abandoned paddy fields. Journal of Rural Planning Association 2, 235-240 (J)

Yamaga S (1971): *Nanboku-danmen kara mita Musashino no toshi-ka* (Die Verstadterung der Musashino-Diluvialplate westlich von Tokyo im Nod-Sud-Profil). Bulletin of Tokyo Gakugei Univ, Series Ⅲ (Social Science) 23, 109-119 (J-G)

Yamamoto S (2001): Studies on the effect of changes in rural landscape structure on secondary forest plants in Japanese rural areas. Bulletin of the National Institute for Agro-environmental Sciences 20: 1-105 (J-E)

Yamanoshita T, Kawazoe T, Tange T, Masumori M, Nuyim T, Kojima K, Urapeepatanapong C, Wasuwanich O, Morikawa, Yagi H, Sasaki S (1999): Adaptation of *Melaleuca cajupti* to flooded conditions in tropical peat swamps. In: Ohga K (ed) Proceedings of the International Symposium for "Can Biological Production Harmonized with Environment?", Tokyo, October, 1999. Asian Natural Environmental Science Center, The University of Tokyo, Tokyo, pp. 469-472

Yamase K (1998): Vegetational changes with the magnitude of improvement cut in secondary forest of *Pinus densiflora*. Journal of the Japanese Institute of Landscape Architecture 61: 567-576 (J-E)

Yamazaki H, Aoki K, Hattori T, Takeda Y (2000): Increase of species diversity due to management of vegetation in satoyama (rural forests). Journal of the Japanese Institute of Landscape Architecture 63: 481-484 (J-E)

Yokohari M, Brown RD, Katoh Y, Yamamoto S (2001): The cooling effect of paddy fields on summertime air temperature in residential Tokyo, Japan. Landscape and Urban Planning 53: 17-27

Yokohari M, Brown RD, Takeuchi K (1994): A framework for the conservation of rural ecological landscapes in the urban fringe area in Japan. Landscape and Urban Planning 29: 103-116

Yokota T(1973): *Tsurumi-gawa ryuuiki ni okeru tochi-riyou henka* (Landuse change in Tsurumi-gawa river watershed). In: Noma S, Yokota S (eds) Urbanization and metamorphosis of existing rural communities in Tsurumi gawa river watershed. *Toshi Kenkyu Houkoku* (Transaction of Urban Research) 36, 5-21 (J)

Yui M (1988): *Mori ni sumu yatyou no seitai* (Ecology of birds living in forests). Soubun, Tokyo (J)

Yui M, Ishii N (1994): *Ringyo to yasei-tyojyu no kyouzon ni mukete: – Shinrin tyojyu no seisoku-kankyou hogo kanri* (Toward the coexistence between forestry and wildlife: – protection and management of habitat for wildlife in forests). Nihon-ringyo-gijyutu-kyokai, Tokyo (J)

Websites

Abe F: http://ss.ffpri.affrc.go.jp/labs/kanko/338-4.html (J)

Environment Agency of Japan (1995): National Strategy of Japan on Biological Diversity. http://www.env.go.jp/en/pol/nsj/index.html (J-E)

FAOSTAT: FAO statistic database. http://www.apps.fao.org/ (J)

Folktale of Okayama: http://www2j.biglobe.ne.jp/~minwa (J)

Takao Communication: Folktales at Takao. http://www.asahi-net.or.jp/~fd4h-krng/ (J)

The Association for Promoting Advanced Science and Technology of Forest Resource Utilization (APAST): http://www.d4.dion.ne.jp/~twra/ (J)

The Association for Promoting Technology of Biomass Energy: http://www.d4.dion.ne.jp/~twra/newpage4.htm (J)

Tomakomai Mill, Oji Paper Co. Ltd: http://www.ojipaper.co.jp/f_m/f02/to_gai.html (J)

Glossary of Terms

Chamaecyparis obtusa – Japanese cypress, *Hinoki* in Japanese. This is a common species of plantation woodlands of Japan. It requires periodic thinning but, in many areas, it has currently lost its economic value and, thus, thinning is not being done. Consequently, the trees are too thin and the roots are too shallow, so the trees are vulnerable to destruction by typhoons and major disasters.

Common land – In traditional Japanese landscapes, common forests were available for anyone to harvest wood for fuel or for charcoal production. In addition, common grasslands were available for the harvesting of food for livestock.

Coppice – A forested area that is used for the production and harvest of fuel wood and charcoal. Trees are periodically felled, but the stumps and roots are left intact so that sprouts emerge and grow for future harvest.

Cryptomeria japonica – Japanese cedar, *Sugi* in Japanese. This is the most common species in plantation woodlands in Japan. These trees are normally harvested between 50 and 100 years after planting. Since World War II, planting of *C. japonica* has taken place on a massive scale in Japan. The *C. japonica* plantations are in need of thinning, but because these woodlands are currently not as economically viable as they once were, in many areas they have been abandoned without management. The trees are consequently too thin for timber and there is a threat of major disaster: typhoons might be able to destroy the forest and/or cause landslides.

Edo period – (1603–1866) The Edo period was the time during which Japan was united, for the first time, into one country. The Tokugawa Shogun Family had political control and the Emperor was a figurehead with no political power. In many areas, the rural landscapes of today were formed during the Edo period. This is an important reference point for understanding the rural landscapes of Japan and, in many cases, the reference for the point in time to which people are trying to restore the landscape.

Jinsoku map – The complete name in Japanese is *jinsoku-sokuzu*, literally a quick survey map. These maps were produced at the beginning of the Meiji period and were the first maps of Japan that followed Western geographical mapping technologies. There were two major roots in geography at the time: French and German. The jinsoku maps were originally based on the French method, which used colours. Surveys were done by the army at the time and cover only the capital region of Japan. The intent of the maps was to provide for defense of the capital region. Because of their nature as quick survey maps, the jinsoku maps are not precise in detail, but some information on land use and vegetation can be extracted using notes and comments on the map.

Jomon period – (10 000–7000 B.C.) The prehistoric period before rice cultivation was introduced into Japan. People lived primarily by hunting and gathering.

Meiji period – (1867–1911) Named after the Emperor of the time, this was a period when the Shogun domination of Japan was terminated and the Imperial Family resumed political power. It is often called the Meiji Revolution. During this time, the country ceased its isolation policy and opened its borders to the Western world. This was the beginning of the modernization of Japan and it was also a time of major change in the rural landscape.

Phyllostachys pubescens – *Mosochiku* in Japanese. This bamboo replaces the red pine (*Pinus densiflora*) as they die from pine wilt disease in western Japan. Introduced from China, mosochiku is a shallow-rooted, fast-growing, and aggressive plant.

Pinus densiflora – Japanese red pine, *Akamatsu* in Japanese. The direct translation is red pine. The most common pine tree in mid- and western Japan, *P. densiflora* is often found on the dry ridges of satoyama landscapes. In the Kanto area, *P. densiflora* is found on relatively dry and nutrient-poor soils. In western Japan, *P. densiflora* is the dominant species of satoyama landscape. Over the past 30 years, pine wilt disease has seriously affected the *P. densiflora* of Japan and a considerable amount of *P. densiflora* forest has been destroyed.

Pleioblastus chino – *Azuma-nezasa* in Japanese. This bamboo species is common, especially in eastern Japan. This species dominates the forest beds of satoyamas in the Kanto region. It is very fast growing and dominating. It is so thick that virtually nothing else can grow with it, so natural succession is halted or, at least, strongly controlled.

Prefecture – A political division that is similar to the provinces in France or Canada. In Japan, much of the political power is with the Central Government.

Quercus arcutissima – *Kunugi* in Japanese, this is a common deciduous oak tree in coppice woodlands in Japan. It is found in mid- and western Japan and has long been used for charcoal and firewood. *Quercus arcutissima* prefers wetter soils than does *Quercus serrata*, so it is mostly found on the lower half of slopes.

Quercus myrsinaefolia – *Shirakashi* in Japanese. This is an evergreen oak tree regarded as the potential natural vegetation for most of the Kanto Plain.

Quercus serrata – *Konara* in Japanese. This is the most common deciduous oak tree in coppice woodlands in Japan. It is most common in mid- and western Japan and has long been used for charcoal and firewood. *Quercus serrata* prefers relatively dry soils, so it is often found on the upper-half of slopes.

Satoyama – From *sato no yama*, which means, literally, low mountains or uplands in or around rural areas. The term describes woodlands of rural Japan that have been coppiced and/or used for the harvesting of litter for agricultural fertilizers.

Satoyama landscape – The traditional landscape system of Japan that was comprised of woodlands that had been coppiced and/or were used for the harvesting of litter for agricultural fertilizers (satoyama), common grasslands, settlement areas, vegetable fields, and paddy fields.

Showa period – (1926–1987) This was a time of massive change in the rural landscape of Japan. Due to a very successful economy, and the fuel and fertilizer revolutions, the landscapes of rural Japan were transformed and largely abandoned. This is the third major change brought to the rural landscapes of Japan. Modern society was formed during this period and it was a major turning point in history for the rural landscapes of Japan.

Terraced rice paddies – In hilly landscapes, the slopes were made into small diked terraces to provide areas for rice to be produced. Rice plants are flooded with water one or more times during their growing season, so the terraces must be leveled and have a source of water.

Yatoda – Local variant of the word yatsuda.

Yatsu – Also known locally as *yato,* these short and narrow valleys cut into the surrounding uplands or hills. There is no clear definition of the requisite length or width of yatsu, but it is generally regarded that the width at the mouth will be less than 200 m and the length will be less than 2 km, with a width that is normally less than 100 m and a length normally less than 1 km. These valleys have been used for rice cultivation at the bottom of the valley and often have an irrigation pond at the top of the valley. The yatsu are often surrounded by woods along the slopes surrounding the valley. These valleys are one of the major components of satoyama landscapes. Here you find very rich wildlife species due to the variety of landscape elements in the area. The water bodies include an irrigation pond with stabilized water of various depths, irrigation canals, and springs along the valleys. Also, very shallow stable water in the paddy fields.

Yatsuda – The '*-da*' stands for paddy fields. So, the term *yatsuda* stands for paddy fields in yatsu valleys.

Yayoi period – (7000–2000 B.P.) This is the time when rice cultivation was introduced to Japan and people started to work on the land. Few records remain, so the landscape patterns are not well known. It is thought that the landscape patterns in rural Japan did not change much between the Yayoi and Edo periods.

Map of Japan
Regions and Prefectures

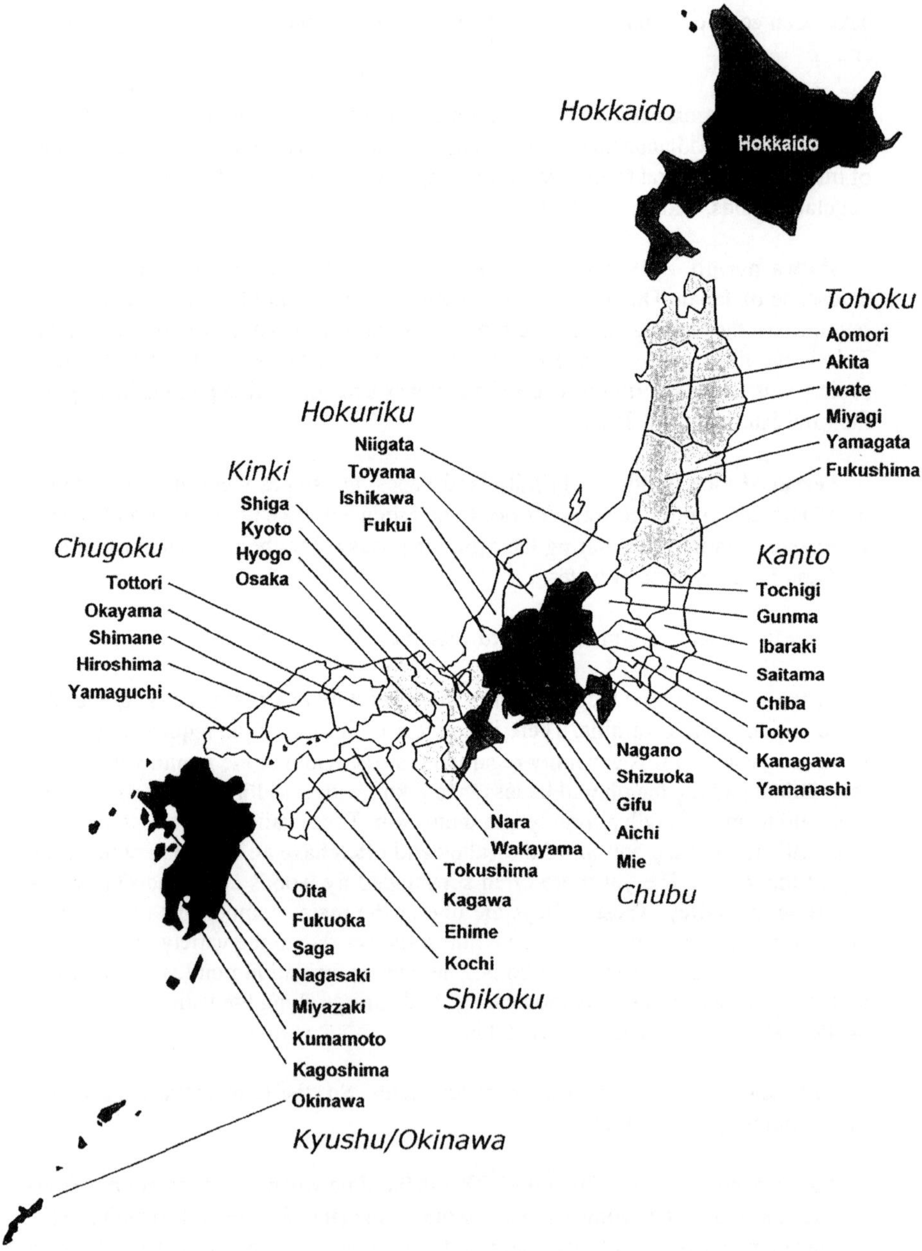

Subject Index

GPSR Compliance

The European Union's (EU) General Product Safety Regulation (GPSR) is a set of rules that requires consumer products to be safe and our obligations to ensure this.

If you have any concerns about our products, you can contact us on ProductSafety@springernature.com

In case Publisher is established outside the EU, the EU authorized representative is:

Springer Nature Customer Service Center GmbH
Europaplatz 3
69115 Heidelberg, Germany

Batch number: 10755292

Printed by Printforce, the Netherlands